Thomas Twining

Science for the People

a Memorandum

Thomas Twining

Science for the People
a Memorandum

ISBN/EAN: 9783744644907

Printed in Europe, USA, Canada, Australia, Japan

Cover: Foto ©berggeist007 / pixelio.de

More available books at **www.hansebooks.com**

SCIENCE FOR THE PEOPLE:

A

MEMORANDUM

ON VARIOUS MEANS FOR PROPAGATING SCIENTIFIC AND PRACTICAL KNOWLEDGE AMONG THE WORKING CLASSES, AND FOR THUS PROMOTING THEIR PHYSICAL, TECHNICAL, AND SOCIAL IMPROVEMENT.

ADDRESSED TO

LORD HENRY GORDON LENNOX, M.P.,

Chairman of the Council of

THE SOCIETY OF ARTS,

BY

THOMAS TWINING,

One of the Vice-Presidents of the Society.

LONDON:
PUBLISHED BY C. GOODMAN, 407, STRAND, W.C.

1870.

SCIENCE FOR THE PEOPLE.

INTRODUCTION.

My Lord,

For more than two years a tedious relapse of a former chest complaint has prevented my joining the meetings of the Council of the Society of Arts; but my thoughts and exertions have been only so much the more concentrated on the various means for facilitating the attainment of scientific and practical knowledge by the people. The lively interest taken in this subject by your Lordship and my other colleagues, induces me to submit to the Council, as on previous occasions, a brief report of my endeavours. I will venture to add a few considerations respecting the general system of Elementary and Industrial Instruction which I conceive to be wanted by the working classes of this country, and in which educational facilities like those I am engaged in organizing on a small scale, might acquire a full development.

Believe me to remain,
My Lord,
Yours respectfully,
T. TWINING.

PERRYN HOUSE,
 TWICKENHAM,
  April, 1870.

CONTENTS.

SECTION 1.—ECONOMIC EXHIBITIONS AND MUSEUMS...Page 9

Gallery of Domestic Economy at the Paris International Exhibition of 1855. The Twickenham Economic Museum. Its range of objects. Definitions. Obstacles to the diffusion of Economic Knowledge. Respective labours of Theology and Science.

SECTION 2.—FREE POPULAR LECTURES ON THE SCIENCE OF COMMON LIFE, ORGANIZED IN CONNECTION WITH THE TWICKENHAM ECONOMIC MUSEUM..................Page 15

Ineffective Science Teaching. First attempt at a new Method. Difficulties to be overcome. Binary mode of delivery adopted for the new Course. Its nature and purpose. Its intended publication. Providing of Illustrations. Working details; including:—distribution of Programmes, &c.; selection of localities; choice of Chairman. Orderly and appreciative audiences. Summary of the results of experience.

SECTION 3.—EXAMINATIONS IN CONNECTION WITH THE POPULAR LECTURES... Page 34

Experiment at the Lambeth Baths. Double purpose of Examinations; viz., to afford a guidance, as well as a test. Science for Working Men. Examination Programme. Plan of operations based on an open Syllabus. Summary of advantages.

SECTION 4.—PRACTICAL BIONOMY OR THE SCIENCE OF COMMON
LIFE, AS A PART OF PRIMARY EDUCATION.......Page 46

The kind of Science required for guidance in Daily Life. Practicability of introducing it as a part of popular Primary Education. Answer to the question, can Children imbibe Science? Importance of an understanding on the use of terms. Adaptability of the proposed plan to various educational schemes. Series of propositions relative to the mode of tuition. Summary of encouraging data. Question of Time. Question of providing Teachers; difficulties which present themselves, and means for overcoming them. Proposal for Examinations in Practical Bionomy.

SECTION 5.—SCIENCE AS A PART OF TECHNICAL INSTRUCTION.
Page 59

Comparison of the knowledge required for bionomic and for technical purposes. Method for classing technical requirements. Their parallelisms and divergencies. The kind of Chemistry wanted by Working Men. Attempt to supply it in 24 Familiar Lessons. The requirements of the Dyer taken as a first example. Other Trades to follow. Appeal for information and advice.

SECTION 6. — THE EDUCATIONAL WANTS OF OUR ARTISANS.
Page 67

Change in the public mind since 1867. How it may best be turned to account. Centralization; formerly an object of mistrust, now an unquestionable desideratum. Analysis of the Educational wants of our Artisans under the following Heads. A. page 72—Primary Education, including moral training; instruction in scientific and practical knowledge; foundation for technical training; manual dexterity and practical cleverness; art training; music and poetry; recreation, exercise and drilling. B. page 85—Apprenticeship; including: prevailing causes of complaint, and remedies suggested by continental example. C. page 88—Evening classes, Art and Science Schools and the like. Science teaching and Science teachers. Measures to be hoped for from a centralized administration. D. page 92 —Educational Materials. Publication under official authority of educational diagrams, technical handbooks, &c.

vii

F. page 95—Science Museums and Art Galleries considered as a means of Popular Instruction. Defects to be avoided. Principles of organization which should prevail. Means for utilizing existing Collections. F. page 105— Technical Examinations and Certificates. The German System as described in " Letters on the Working Classes of Nassau." Its defects. How these might be avoided, and its best features safely adopted in this Country. G. page 110—The Central Industrial College or Pantechnium. Necessity for such an Institution. How it should be constituted, and where located. H. page 115—The Educational Scheme of the Future. Precautions to be observed, Measures to be initiated, and prospective results to be secured.

Appendix No. 1—Synopsis of the Twickenham Economic Museum.

Appendix No. 2.—Programme of Popular Lectures. (See Section 2.)

Appendix No. 3.— Sketch of a Museum for East London (See Section 6. E.)

Appendix No. 4.—Suggestions for a District Museum. (See Section 6. E.)

Appendix No. 5.—Supplementary Notes, including the following subjects:— A Pan-athenæum of Science and Art. Hygiene in the Medical Curriculum. Prize Examinations in Hygiene. A Manual of Hygiene. Two Lectures on Health and comfort for Seamen, and a fuller Course of Nautical Hygiene for Commanders and Officers in the Merchant Service.

SCIENCE FOR THE PEOPLE.

SECTION I.

ECONOMIC EXHIBITIONS AND MUSEUMS.

I HAD not long been a member of the Committee of the Labourers' Friend Society, which I joined in 1847, before the insight which I had the opportunity of acquiring into the condition and resources of the working men of this country, collated as it were with the knowledge of the working men of other countries previously acquired during a long sojourn on the continent, led me to the conclusion that much benefit might accrue to both from an interchange of notions, habits, and contrivances; but that still greater advantages might be derived from bringing the united influence of science and inventive industry, to bear directly and constantly on the requirements of the million, and by spreading everywhere a knowledge and an appreciation of the results thus obtained.

These considerations, submitted to the Council of the Society of Arts in 1852, and more fully developed in a Memorandum addressed in March, 1855, to our then Chairman, LORD EBRINGTON,* were through the influence of the VICOMTE DE MELUN, embodied in a special resolution by the Philanthropic

* Now EARL FORTESCUE.

Conference held at Paris in July of the same year* A Committee was appointed which, thanks to the very friendly assistance afforded by M. LE PLAY, was soon enabled to organize as an annex to the Universal Exhibition at the Palais de l'Industrie, the Gallery of Domestic Economy, of which some of my colleagues will remember the inauguration on the 15th September, in the presence of a numerous concourse of members of the Society of Arts. The principle thus initiated,** gained everywhere such ready acceptance, that nearly every international Exhibition held on the continent since that time, has devoted a department to means for the improvement of the condition of the Working Classes, and that several important Exhibitions have been held in various countries for that distinct purpose.

Whilst however temporary exhibitions of this kind are, when well directed, of immense advantage for giving to manufacturing industry an impulse in the required direction, and for inducing the spontaneous juxtaposition of articles which it would be otherwise very difficult to get an opportunity of comparing with each other, to say nothing of the economic gems thus rescued from obscurity, yet I have felt all along that the most practically useful lessons would be those taught by permanent collections, organized on more strictly educational principles, so that one might not only see the things to be adopted or eschewed, but

---

* "La Réunion émet le vœu, conformément aux vues exposées dans le mémoire de M. Twining, qu'il soit constitué dans les divers pays un musée économique permanent, où seront réunis et classés tous les articles destinés à l'usage domestique et qui se distinguent par des qualités d'utilité, de solidité, et de bon marché, ainsi que les procédés et les appareils qui se rapportent à l'hygiène et à l'assainissement des habitations, des ateliers, etc."

** It was proclaimed as follows in a Report on this special Exhibition prepared on behalf of the Committee by M. Augustin Cochin:—
"Désormais, aucune Exposition universelle ne doit avoir lieu sans qu'un large espace soit réservé à l'exhibition spéciale des objets utiles au bien-être physique ou au développement intellectuel des classes les plus nombreuses de la société."

learn at the same time the "reason why." It was in this spirit that I began in 1856 to form the permanent and educational Exhibition of the things appertaining to Domestic and Sanitary Economy, which from its being devoted to the furtherance of what may be called ECONOMIC KNOWLEDGE, has taken the name of ECONOMIC MUSEUM.

It is to the various modes of diffusing this ECONOMIC KNOWLEDGE among the People, that the following pages will be chiefly devoted. To attempt to compass it with a precise definition, would be to deprive it of the elasticity of circumscription which enables it to promote man's physical well-being under the most varied circumstances and conditions of life; but it may be briefly said to embrace in this essentially utilitarian direction, everything that everyone would say that everybody ought to know. Thus it is of unquestionable importance for all classes of society, and especially for those whose income is small, to know how their dwellings should be constructed in accordance with sanitary principles; what household improvements they may derive from the discoveries of science, or borrow from the customs and appliances of other nations; what fabrics they should wear; what food they should eat, and how it ought to be cooked; how they may distinguish things which are genuine, wholesome, substantial, durable, and really cheap, from those which are cheap only in appearance, and, in short, how they may live with judgment, and get the best money's worth for their money.

To these elements of comfort should be added in a comprehensive interpretation of Economic Knowledge, not only the most practical means for the avoidance of harm or injury and the alleviation of suffering, but also that amount of information concerning articles constantly used or seen, to lack which would be palpable ignorance.*

* For more definite indications, see the synopsis of the chief series of Illustrations in the Twickenham Economic Museum, annexed to this Memorandum as appendix No. 1, together with a few explanatory remarks

It is self-evident that the study of such a range of subjects, embodying as they do in a more or less direct manner, the applications of scientific facts and principles to the concerns of daily life, presupposes a preparatory acquaintance with these facts and principles on the part of the earnest student. On looking closer, we find that this indispensable foundation of his economic studies, consists mainly of the elements of Physics, Chemistry, and Human Physiology, with certain general notions of Natural History.

To make it more clear that ECONOMIC KNOWLEDGE ought to include these scientific elements as well as their practical applications, I sometimes call it ECONOMIC SCIENCE, or substitute the more comprehensive expression of SCIENCE OF COMMON LIFE; but a title which I shall take the liberty of adopting more frequently, having obtained for it the sanction of eminent scientific friends, is PRACTICAL BIONOMY, which indicates more clearly the union of Science and Common Sense for our practical guidance in Daily Life.

What I have found to be the greatest bar to the diffusion of sound principles of Domestic and Sanitary Economy, is the almost total absence of the above preparatory knowledge, nay of all Scientific training among the bulk of the Community at large and the consequent want of ability on their part to enter into the rationale of the merits or defects either of the things now in use, or of those proposed as substitutes. An artisan and his wife visiting the admirable Food Department of the South Kensington Museum, may be struck and interested amazingly by some of the sensational illustrations and labels, but they are so much at sea in all that relates to the chemistry of nutrition, that they would scarcely venture to alter one item in their daily fare in accordance with a scientific dietary. They feel indeed as

taken likewise from the Museum Programme. Copies of the latter and other papers relating to the educational movement of which the Museum is the leading feature, may be had on application to the Secretary, William Hudson, Esq., Economic Museum, Twickenham.

would feel many a classical scholar if he were invited to ramble through field or forest with a botanical work on the Fungi, and to feast on a variety of mushrooms he had never touched before. He would thank you for the suggestion, but prefer sending for the old article to the old shop.

I am sorry to say that the difficulty of inducing a due appreciation of the value of science in unscientific matters, has met me in all directions. I have found it among schoolmasters and even among clergymen, on whom I had particularly reckoned for propagating among the poor, intellectual means for physical improvement, and among the rich, notions of judicious and discriminating benevolence. It is true that from some of the more enlightened members of the clergy I have received the most gratifying tokens of sympathy and support. I refer with particular pleasure to an encouraging letter received from his Grace the present ARCHBISHOP OF CANTERBURY, then Bishop of London, respecting a small pamphlet by JOHN BERRILL which appeared in 1864, under the title "The Christian Teacher's visit to the Twickenham Economic Museum." I could cite other examples of distinguished divines of various denominations concurring in the conviction that the frame of mind suited for imbibing spiritual truths, is far more likely to be found among the intelligent inmates of a healthy home, than among those whom ignorance has degraded to a torpid state of misery. Unfortunately there are many churchmen who appear not to feel the force of this principle; or if they concur in the theory, they seem not to consider themselves the most fitting instruments for carrying it out in practice; they prefer leaving to secular benevolence to provide for secular wants, and raise a boundary between the vineyard where theology labours for the salvation of the soul, and the open field where science labours for the welfare of the body; partly perhaps because science was not in their College curriculum, and partly because science in general is apt to be affected in their minds by the anathemas which the church has pronounced against certain departments of science in particular.

I must leave for another opportunity the consideration of the various means by which churchmen and *savants* might be brought to an *entente cordiale* honourable to both and infinitely profitable to the community; but in the meantime I shall have done something towards promoting this happy result, if I can show science setting aside controversial theories, to dispense the most wholesome and humanizing knowledge, where ignorance is engendering misery and callousness; and opening the way alike to superstition and infidelity.*

* But for the deplorable want of that enlightenment which science bestows, we should not see the so-called "Peculiar People" carrying into actual practice the perversion of Scripture Texts, nor would it pay to publish books like that which was some months since largely placarded under the title of "Christ is Coming." The following extract concerning the deluge will give some idea of the level at which science and common sense stand in the mind of its author:—

"Understand, oh man! the round earth has a bulk about forty-nine times greater than that of the moon, and therefore is more saturated with solar heat than is the moon. The moon contains less solar heat, in proportion to its bulk, than does the earth. Therefore understand, oh man! that when the Lord God placed the moon much nearer to the round earth; the ground and waters of the earth, the dormant solar heat within the round earth and the moon, became together, as it were, a great and powerful galvanic battery; for the dormant solar heat which solidified the ground, quickly loosened and sped towards the moon, the round earth becoming, as it were, a sun to the moon, and the whole surface of the ground became quickly split and disrupted, and great chasms were made in the shallow oceans. Great clouds of vapour arose from the fast decomposing ground and the fast decomposing mountains, during the time the moon was nearer to the round earth, which turned into rain in quantity sufficient to cover to a great depth with water the whole surface of the round earth—the waves rolling over the highest mountains. The solid surface of the round earth and the solid surface of the mountains were converted into water, and the mountains dwindled greatly.

The Lord God, after forty days and forty nights, removed back the moon, which caused the round earth to be no longer, as it were, a sun to the moon, and decomposition of the ground, and of the mountains ceased; and gradually the oxygen and hydrogen of the superfluous waters entered into fresh combinations with the only other earthly element, nitrogen; the solar heat again became dormant, and they solidified once more into earthy and metallic matter, which settled, with the sweepings of the oceans and of the old continents and islands to the bottom of the water, into strata. The waters were many days shrinking into the comparatively small quantity that now forms the waters of the round earth."

SECTION II.

FREE POPULAR LECTURES
ON
THE SCIENCE OF COMMON LIFE,

Organized in connection with the Twickenham Economic Museum.

Having in 1861 removed my Economic Collection to a building erected in the immediate proximity of my residence, I was, notwithstanding my infirm state of health, so far enabled to develop it in about three years, that I thought I might begin to organize in connection with it a series of popular lectures of corresponding scope and classification, preceded by an introductory explanation of the most indispensable rudiments of scientific knowledge; those rudiments for the want of which my museum had for many visitors been almost a dead letter. The first question however to be solved, was whether the working classes could be induced to listen in earnest to a series of methodical lessons on subjects not of the kind commonly considered attractive. Much had been done to give the audiences at Mechanics' Institutes a false and unfavorable idea of Science. When genuine and educational, it had often been in substance too high and dry, and in form too didactic and technical; so as to require on the part of the artisan an amount of preparatory knowledge which he could scarcely be expected to possess, and which indeed it seemed to be nobody's business to give him. When amusing, it was rendered so by experiments more calculated to be admired than understood, and it consisted mainly of sensational bits, picked here and there from the range of scientific knowledge, and which without the general context could scarcely produce the clear impressions required for practical use.

It was to test the possibility of remedying this state of things, that I composed in 1864 a detailed explanatory Syllabus of six connected Lectures respectively entitled as follows :—

1. "The Alphabet of the Science of Common Life; or, a first peep into the mysteries of Health and Comfort."
2. "A good Home, and what belongs to it."
3. "Furniture and Clothing; and Health as affected by them."
4. "Food: its purposes, principles and resources. How to make meals palatable, wholesome and cheap. Beverages."
5. "Fire: what it is and how to make the best of it. Contrivances for Ventilation."
6. "Good Health, and how to keep it."

Each lecture thus sketched out was provided with an ample assortment of specimens, diagrams and apparatus, prepared and suitably packed for circulation at the Twickenham Museum. Through the valuable co-operation of the Ladies' Sanitary Association, and the exertions of Mr. SALES, then Secretary of the Metropolitan Association for the Instruction of Adults, arrangements were made for the delivery of the Course at ten places of Meeting in the Metropolis during the ensuing winter. The six lectures were entrusted to three competent persons, who severally undertook to adhere as closely as might be to the substance of the Syllabus, clothing it in their own language, and who acquitted themselves of the task in a very creditable manner.

The result was in the main highly satisfactory; proving that even illiterate audiences could imbibe the more practical portions of science with an earnest appreciation; and that the plan of preceding the study of Domestic Economy with an introductory summary of the sciences on which it depends, was decidedly a step in the right direction; but at the same time it became obvious that the *modus operandi* which had been adopted, was susceptible of considerable improvement. Firstly, the amount of matter, especially as regards the first lecture, was far too great for the limits assigned to it, and seemed indeed to be bursting them in all directions. Secondly, my syllabus, however detailed and explicit it might be, could not sufficiently control the selection of the subjects, and still less secure a uniform treatment of the whole range of them by different lecturers. Thirdly, I found that centralization and unity of action were as necessary for the success of my small educational experiment, as I imagine they will be found to be in our national educational system.

Further experiments subsequently undertaken in pursuance of these conclusions, confirmed and extended them. It became more and more evident that to deliver a fluent and taking lecture on materials prescribed by me, and in accordance with my particular views, required consummate abilities which a lecturer possessing them would not like to have thus fettered. The abundance of his knowledge would at times be thoughtlessly poured out to the detriment of the particular facts which I had thoughtfully selected; the ardour of his mind would ever and anon break the line of connected purposes, and carry him off at a tangent into the realms of speculative fancy, his facile eloquence would dilate on favorite themes, and

B

three quarters of the allotted time would be consumed before one half of the allotted matter was got through; so that the remaining half would either be cut and maimed, or galloped over at a headlong pace.

I enter into these details, because the same difficulties will have to be guarded against by all persons or bodies acting as centres of instruction, and undertaking to radiate it to a distance. It is true that those invested with governmental authority, or otherwise endowed with ample resources, might be able to train in process of time, a staff of special professors up to the mark in the spirit of their mission as well as in the ability to fulfil it; but generally speaking, uniformity of tuition is exceedingly difficult to secure where any freedom of action is allowed. As regards my own case, as I had neither the strength nor the inclination to become a lecturer myself, there was no alternative but that the proposed Course should be written out *in extenso;* and having failed in an attempt to get the authorship into better hands, I was obliged to become author myself. I accordingly prepared by the autumn of 1866, under the title of "Science made Easy," the full text of five familiar lectures on the Elements of Physics and Chemistry as part of the *preparatory portion* or *scientific groundwork* of a comprehensive course on the SCIENCE OF COMMON LIFE. In doing this I endeavoured to carry out to the best of my power the instructions which I had prepared for others, as to the manner in which the subject matter should be selected and arranged, and as to the style in which it should be expressed.

Here however I was met by another ominous question: would Working Men listen to a *read* lecture? As a rule the reading of a discourse of any kind only answers when there is nothing to do but to read, and is a failure when the delivery of the text is broken by experiments, writing on the Black-board, or any other kind of visual demonstration. The expedient for overcoming this difficulty which happily occurred to me, was very simple, but proved so effective that I ascribe to it in great measure the uniform success which has attended the delivery of my Course. It consists in the joint action of a Reader and a Demonstrator. Wherever a specimen is to be shown, a Diagram to be pointed to, or an experiment to be performed, a cross (X) in the text warns the Reader to make any pause that may be required. The Demonstrator who has before him a full list of instructions, with every device for enabling him to be ready at the right moment, does the needful, and the reading is resumed without the least embarrassment or loss of time. I am the more induced to lay some stress on the remarkable success which this plan has obtained, because I believe it to be a new one, and to be susceptible of rendering notable service to the cause of popular instruction. A Reader cannot be expected to deliver a written or printed lecture satisfactorily, if he has to go to and fro between his text and his diagrams, or his apparatus. On the other hand a professor who can deliver a whole educational course in a concise yet easy style, without further guidance than a few notes, and who at the same time is a skilful experimentalist, is an expensive luxury, even in London, and almost unobtainable in most provincial localities. But on the contrary there is scarcely a country town where the vicarage or the school cannot supply a good Reader for a philanthropic subject, whilst the Doctor or the Chemist of the place will be sufficiently up to the performing of any amount of chemical or other experiments, involved in expounding the Science of Common Life.

It is true that it will not always be easy to find a Reader so capable of captivating a working class audience by the power and flexibility of his voice and the popular style of his delivery, as the one I have had the good fortune to find in Mr. W. FREEMAN, the Curator of my Museum; nor will a Demonstrator always be at hand so clever at manipulating, and at the same time so competent to answer any scientific questions that may be put to him at the close of a lecture, as my Secretary, Mr. W. HUDSON. The peculiar smoothness and homogeneity with which their joint action progresses as rapidly as that of the most fluent and experienced extempore professor, will not be attained to by many provincial amateurs, but neither will this degree of efficiency be expected by provincial audiences, or required for making this mode of teaching Science a welcome element in Penny Readings, and a valuable resource for Schools.

For succinct indications of the nature and purpose of my Course, I cannot do better than refer to a Programme which has been abundantly printed and distributed in four pages quarto, and of which the substance is given as Appendix No 2 to this Memorandum. Its heading "SCIENCE MADE EASY," followed by the explanatory title "FAMILIAR LECTURES on the APPLICATIONS OF SCIENCE TO THE REQUIREMENTS OF DAILY LIFE, offered gratuitously to Institutions established for the promotion of POPULAR IMPROVEMENT;" then the further explanation that "This Course of Lectures is intended to unite in an entertaining form, the various departments of practical knowledge which tend to the promotion of HEALTH and COMFORT," and further down the proviso, "That the public be admitted FREE, the RECREATION and PRACTICAL BENEFIT of the WORKING CLASSES being the sole object in view;" all this marks at once the purpose of the Course, and the character of the audiences chiefly intended to be addressed. A condensed syllabus shows the contents of the nine Lectures of which the Course at present consists; three being allotted to select elementary portions of Mechanical Physics, and similarly one to Chemical Physics; one each to Inorganic and Organic Chemistry; one to outlines of Natural History, and two to Human Anatomy and Physiology.

This elementary knowledge, carefully selected as the most indispensable for understanding the rationale of Daily Life, forms a substantial foundation or groundwork, on which I hope, should my health allow, to erect a superstructure embodying the practical applications of Science to the various departments of Household and Health Economy, and consequently including the following subjects, as shown by an announcement at the end of the Programme:— "Dwellings as they should be, and the art of constructing them. Building Materials. Fixtures, Furniture and Household Utensils. Textile Materials. Fabrics, Dress. Food. Warming, Lighting, and Cleaning. Public and Personal Hygiene. Safety from injury, and means of relief, &c."

Thus this Course (which I will briefly designate as my Popular Course, to distinguish it from a rather more advanced Course for industrial purposes to which I shall have occasion to allude in another Section) is intended to consist ultimately of two parts, devoted respectively to elementary and applied Science. I may remark however that on the one hand the examples selected to illustrate elementary scientific principles, are of the most practical and every-day character, that on the other hand the review of the more tangible requirements of Daily Life, will be studded with scientific facts and experiments reserved for the purpose; and that there will indeed be throughout, such a concatenation of ideas as to unite virtually the whole, and present a continuous and methodical *exposé* of the Science of Common Life, or Practical Bionomy, specially adapted to the requirements of the million, and brought as much as possible to that complexion which, without any sacrifice of sound teaching, may justify the title of "Science made Easy."

I think of proceeding at once to the publication of the present Nine Lectures in a bold type and at as cheap a price as possible; giving every information that may facilitate their use by provincial Schools and Institutes. The greatest difficulty with which they will have to contend, will be the providing each Lecture with a numerous set of Illustrations similar to mine. It is true my Models and Apparatus are purposely made of common materials, and are reduced to the very simplest contrivance, but even this simplicity is in many instances the result of much labour and experience; nor would a provincial beginner readily hit upon the ingenious devices by which Mr. FREEMAN now packs safely in a single box or trunk of moderate dimensions, the incongruous requirements of a comprehensive Chemical Lecture. Should my Course, when printed, find sufficiently extensive favour, I would endeavour to induce a well-known dealer in educational appliances, to prepare analogous illustrations at cheap rates, packed ready for travelling. One of the greatest obstacles he would have to overcome, would be the deficiency of the stock of Diagrams hitherto published, which has obliged me in a great many instances to have recourse to hand-made drawings.

The following account of various working details of my lecture scheme, may be useful to others in organizing similar attempts.—The Programme in four pages quarto, mentioned above, is kept in type during the whole of the lecturing season, and copies printed on tinted paper are freely supplied, generally by 500 at a time, to the Institutions to which the Course has been promised. They similarly receive gratis for distribution, any number they may require of special hand-bills giving the dates of the lectures.*
The Managers of the several Institutions generally undertake at their own expense, the printing and posting up of large Placards.

* Specimens of these as well as of all other papers relating to the Lectures will be supplied on application to the Secretary of the Twickenham Economic Museum.

All these subsidiary matters are by no means to be disregarded; nor does it generally answer to give the Lectures where there is not some intelligent and active person on whom reliance can be placed for superintending the machinery of publicity. They succeed best where an energetic minister or missionary, has induced habitual gatherings of the people in large numbers, by mixing innocent recreation with religious and secular instruction. There my Lectures are quite at home; for it has been my constant endeavour to leaven them with a spirit of Christian morality, and to show how religion and science may, and always should, go hand in hand.

The most numerous audiences have been those at the Lambeth Baths, where my Course was inaugurated on the 16th November, 1866, and has lately been delivered for the fourth time. I must explain that the vast Hall of that Establishment, which serves as a Swimming Bath in summer time, is rendered perfectly dry in winter, when it is fitted up so as to form a most eligible place of resort for the Working Classes of the neighbourhood. Through the liberality of Mr. SAMUEL MORLEY, M.P., it is yearly placed at the disposal of the Rev. Mr. MURPHY, who occasionally makes use of this noble space for one of those Working Men's Industrial Exhibitions which he has so successfully initiated, but otherwise appropriates it to a continuous weekly rotation of devices for uniting instruction and amusement.

I feel likewise that it is my duty to mention the Evangelist's Tabernacle in Golden Lane, where similar results are being achieved through the highly disinterested and praiseworthy exertions of Mr. W. ORSMAN, one of the *employés* of the General Post Office.

Through the recommendation of MISS GRIFFITHS, Secretary of the Ladies' Sanitary Association, I have more recently become acquainted with another place of meeting, on not quite so large a scale, but where similar principles of action have for many years been producing analogous results under the intelligent management of a special Missionary, Mr. T. BEAUMONT. It is the Boatmen's Institute, Paddington, now open indiscriminately to the Working Men of that neighbourhood. There, in a Hall built for the accommodation of about 350 persons, and supplied with every convenience for lecturing purposes, my Course has lately been given to audiences unsurpassed in steadiness of attention.

I feel pleasure in testifying my appreciation of the good effected in a similar manner by the humble but equally effective exertions of Mr. HARRIS, the City Missionary in Love Lane, Shadwell, one of the poorest districts in London, where nevertheless my free Lectures have been listened to for three seasons with the most gratifying earnestness.

As a rule, I have found, as might be expected, that audiences consisting of men and women of the lower social strata, such as are called in by the invitation of *open doors* in a poor locality, require to be presided over by a person who, besides being a good man of business, possesses the art of making them feel at their ease; whereas, a kind of reverential awe seems sometimes to chill their interest, and check the liveliness of their applause, when the chair is filled by anyone to whom they look up as to a condescending patron. This is a light in which I am sorry to say they too often regard the Ministers of our Church. There have been, however, notable exceptions, and moreover, it must be understood that the impressions which I now report are only those which have been reported to me; for my health has not allowed me to be personally present at a single delivery of a Lecture in London.

The ability to preside at a Meeting, large or small, is not a privilege reserved for gentle birth and classical education. At one of the places where my Course was given in the first season, the chair was taken by one who neither stumbled over his own words, nor treated the Meeting with a flow of eloquence when it was not wanted, but could always, when required, speak cleverly and to the purpose, and conduct an evening's proceedings with tact and courtesy. Both he and the audience showed more than once that they were not novices in parliamentary routine. Now the place of Meeting in question, was the Hall of the Old Pye Street Working Men's Club, instituted through the generous and enlightened exertions of MISS ADELINE COOPER, and occupying the Ground Floor of a Model Lodging House, adapted to the requirements of one of the poorest localities in Westminster. The benches were compactly filled with Hawkers, Costermongers, and Labourers, and their clever and efficient Chairman was one who earned an honest livelihood by uniting the trades of Glass and China Mender and Knife and Scissors Grinder.—To give an idea of the discriminating eagerness with which Working Men hail the advent of knowledge, provided it be of the practical kind they require, I will mention that for some time Science had got into evil odour in Old Pye Street, through causes of the usual kind. I accordingly resolved to give MISS COOPER'S friends only one Lecture by way of experiment, and Lecture IV was selected, as being of fair average interest. They were so well satisfied, that they not only asked to have the whole Course, but desired that the Lecture already given, should be repeated in its proper place.*

But more conclusive than could be any isolated instance, is the general tone of orderly earnestness with which the Course has been attended throughout

* I have latterly had several times occasion to test the capabilities of new places of Meeting by tentative Lectures, before deciding as to whether they should have the whole Course. Lecture IV on Chemical Physics, and Lectures V and VI on Inorganic and Organic Chemistry, are found the most eligible for trials of one, two, or three Lectures.

the four seasons of its London career, from Paddington to Holloway and Spitalfields, from Westminster to Shadwell and Stratford, and from Nine Elms and Lambeth to Woolwich Arsenal. The following extract from a letter which I addressed after the conclusion of the first season at the Lambeth Baths, to the Secretary of the Labourers' Friend Society, may help to dispel the doubts which still seem to prevail in some quarters, as to the capabilities of the Working Classes for rational enjoyment:—

"Imagine a miscellaneous assembly, admitted with open doors, in a neighbourhood which used to be looked upon as one of the lowest in the scale of refinement, listening to a regular Course of Scientific Lectures, with so earnest and orderly an attention, that, except when the silence was broken by hearty applause or well-timed manifestations of hilarity, you might, to cite an expression used on the occasion, 'have heard a mouse tap its tail against the wainscot,' and you will fully agree with me, firstly, that great credit is due to those who with untiring Christian zeal have propagated humanizing influences where they were so much wanted, and secondly, that there are in our working population, sterling qualities of great promise, germs of thoughtful improvement, which only want judicious fostering and disinterested guidance, to produce results of infinite value for their physical and social welfare."

These conclusions have been most satisfactorily confirmed by the writer of an article which appeared in the "Daily News" of the 5th of November, 1868, and of which the following is an abridgement. Under the title of "Science among the Costermongers," it describes a visit to Mr. Orsman's Mission Hall, in Golden Lane, where my Course was then being given for the second, and has this year been given for the third time:—

"On Tuesday evening last, a person passing down Golden Lane, a long, narrow, and poverty-smitten thoroughfare, leading from Barbican into Old Street, might have observed numerous individuals belonging, as their costume unmistakeably indicated, to the Costermonger class, silently making their way, amid piles of empty barrows and heaps of decaying vegetable refuse, towards a building, not very pretentious in its external appearance, situated at the rear of the City Baths. The locality is not a very inviting one. In the surrounding extensive labyrinth of narrow and muddy courts and lanes, reside an immense number of street dealers in fruit, vegetables, and other articles, whose efforts to gain a livelihood by the sale of their various commodities, is constantly bringing them into collision with the Metropolitan and City Police. * * * Low lodging-houses, tally-shops, beer-houses, gin-palaces, and small coal-sheds constitute the chief features of the neighbourhood, which for years, had enjoyed the unenviable reputation of being one of the great metropolitan moral wastes. * * * A few years since, it was nearly impossible for a decently attired person to have proceeded through Golden Lane after dark, without being exposed to the danger of insult, or even violence. Now it is otherwise; the numerous ameliorative, social, and religious agencies, busily at work in the neighbourhood, having tended to produce this beneficial effect. Entering the building above mentioned, and which rejoices in the somewhat puritanical appellation of the "Evangelist's Tabernacle," a curious and suggestive spectacle met the eye. In a large room having a spacious platform at one end, and encircled by a strong and commodious gallery, were crowded together some 400 or 500 men, women, and children, belonging, for the most part, to the poorest classes. Many, perhaps the majority, were members of the street-trading community, the rest of the audience consisting of labourers, artisans, workmen's wives, factory girls, shop boys, street arabs, and the like. And for what purpose were they thus voluntarily collected together? * * * They had met for the purpose of listening to a Scientific Lecture on Chemistry. * * * * Painfully conspicuous amongst the audience were often to be seen the cold, passionless features so common among the frequenters of the penny-gaff, or the public-house concert room; but for once, the naturally dull countenances were lit up with a ray of intelligence, as they endeavoured to comprehend the various explanations offered by the Lecturer. * * * * Certainly it was a most suggestive circumstance, to find such a large body of people belonging to a class for whose special edification thousands of illustrated police newspapers, and serials filled with tales of highwaymen, thieves, and murderers, are weekly published, eagerly listening to a Lecture, in which were explained the various properties of Oxygen, Hydrogen, Carbon, Sulphur, and other non-metallic Elements; the phenomena of combustion; the decomposition and recomposition of Water, and the like. * * * * In previous Lectures, those present had had explained to them the conditions of matter, laws of gravitation, mechanics, aerostatics, and hydrostatics, light, heat, and other elementary portions of Scientific Knowledge; but instead of becoming wearied with the formidable mass of technical teaching with which they were threatened, the number of hearers was found to increase considerably with each successive Lecture. The order maintained was admirable." * * * * *

As I sincerely wish in giving the gratifying results of my own experience to stimulate and assist others to undertake in a similar way the diffusion of practical Science among the Million, I will beg leave to address a few further items of advice to the persons who may be that way inclined.

1. When you intend to call in a promiscuous multitude from the streets, and to treat them with a course of scientific instruction, you must do your utmost to make the heaven-born muse of Science leave for a time the clouds among which she is wont to recline in rapturous self-contemplation, casting now and then a look of pity and contempt on all below. You must absolutely induce her to come down to this nether world, to put on a clean apron, and enter a working family's dwelling, to ventilate it and make it wholesome and comfortable, to inspect the furniture and wardrobe, the kitchen utensils and the contents of the larder, nay actually to light the fire and cook a model meal, not forgetting the care of the young ones and of the sick person in the next room. Now if you confidently assure the Working Classes that Science can do all this to perfection, and be a saving instead of an expense, you may be sure that they will give her a hearty welcome.

2. I have found no difficulty in making Working People clearly see, that in order to understand what Science has to say about the concerns of Daily Life, they must first take the trouble to make themselves acquainted with a few indispensable scientific facts and expressions. They readily undertake this, provided they be explicitly assured (and of course every pledge given must be conscientiously redeemed) that all unnecessary technicalities will be avoided, and all difficulties smoothed down as much as possible; that there will be abundance of specimens, models, diagrams, experiments and other devices for making the senses act as helpmates to the memory, and that through all these contrivances, the preparatory and more strictly scientific part of their studies, will be rendered as entertaining as the subsequent economic part, in which the several departments of Domestic and Sanitary Economy will be successively reviewed.

3. One of the chief difficulties to be encountered in selecting the subjects of these two necessary divisions of any methodical course of Practical Bionomy, is the overwhelming abundance of the matter as compared with the limits beyond which common sense forbids us to reckon on the regular weekly attendance of a large popular audience. This difficulty can only be overcome by great pains and discrimination bestowed on the selection of fundamental facts and typical illustrations, by studying the manner in which they can be best and most closely fitted together, and by endeavouring in repeated revisions to condense the matter of many pages into a few, without squeezing out the juice of the subject, and rendering it dry and unpalatable.

4. The necessity of dividing the subject matter into tolerably equal Lectures, so that each of these may take from an hour to an hour and a quarter in the delivery, cannot without considerable trouble be kept from interfering with the natural divisions, and renders more arduous the always difficult task of allotting to each portion and sub-portion of a subject, an amount of attention proportionate to its relative importance. The best remedy is to reserve certain items of elementary Science, especially of Chemistry, for the second division of the Course; allotting them to the several departments of Household Economy to which they specially apply, and where they will afford a welcome variety.

5. One can scarcely be too fastidious as to the accuracy of every statement, or too patient and persevering in the confronting of different Authors on points on which one has not a direct and personal certainty. The discrepancies between various *standard* books on the same subject, and even sometimes between one part and another of a work of high authority, are only known to those who have subjected these matters to scrupulous research.

6. In proportion as Practical Bionomy surpasses many other branches of applied Science in the actuality of its bearings, the grievous multiplicity of the errors, abuses, and frauds on which it throws a detective light, and the radical nature of the reforms which it might seem to suggest, so also, is proportionate prudence required in inveighing against existing prejudices, and in running foul of existing interests. The Economic Teacher of the Working Classes should not so much tell them what to buy and where to get it, as give them the insight which will enable them to judge for themselves; not so much hurl offensive epithets, however well deserved, against those who live by adulteration and fraud, as sow knowledge that will make their present dealings unprofitable, and at the same time, open their way to better ones.

7. Moderation, and a discreet adaptation of precepts to circumstances, are nowhere more necessary than in matters of Hygiene. Exaggerated remedies generate reaction, or replace one evil by another. Systems that work admirably in the combination of circumstances which some countries afford, prove lame and unprofitable elsewhere, and contrivances that have been marvels of success under the management of the inventors or of their intelligent friends, may come to grief in the hands of ignorance and prejudice.

8. Unity of purpose throughout the Course, and the occasional connecting of one department with another by means of mutual references, should bind the heterogeneous subjects of which the Science of Common Life is composed, into a harmonious whole, pervaded from its leading features down to its merest details, with a methodical spirit of forethought and classification. It is a great mistake to deal lightly with considerations of this kind, in writing for the uneducated. METHOD assists both the intelligence and the memory, and the less cultivated and clear the mind to be taught, the clearer should be the teaching.

Great care should be taken to explain all hard words the first time that they are used, or so to construct the phrase as to make their sense self-evident, also, (and this applies particularly to the Chemical Departments) to avoid mixing the names of substances not yet described in the account of those in hand. There should be, as much as possible, a gradual and logical progression from the simple to the complex, from the easy to the difficult; a development of information, step by step, in that connected sequence which makes little children remember so well the Nursery Tale of the "House that Jack built."

9. Rigidity of principle in the selection and arrangement of the subject matter, does not by any means imply rigidity of style. The character of the language which suits a Scientific Discourse, varies immensely, according to the subject and the purpose. It should be solemn where the object is to raise the mind to a true conception of the power and beneficence of an all-wise Creator; it should be sedate and didactic when useful physiological facts and advice can thus be more appropriately imparted; but the general tone to adopt in teaching Working Men the practical difference between Knowledge and Ignorance, is decidedly a cheerful and colloquial one, and in many cases ridicule will be found the best weapon to use against an absurd contempt of the Laws of Nature, and a silly subserviency to those of Fashion. But what as much as anything will secure for wholesome truths a ready and sincere acceptance, is their being offered in a spirit so plainly fraught with sincere Christian benevolence, as to be above all suspicion of an interested motive.

10. My experience of four seasons is decidedly in favour of FREE ADMISSIONS. On two or three occasions I have, in order to conform to the practice adopted at certain Institutions, or by way of experiment, consented to a charge being made for reserved

seats, or to the levying of a penny on non-members; but the result has been invariably unsatisfactory. The amounts obtained by this means have been insignificant, the attendance has been diminished, and as for orderly behaviour, that of the Free Audiences could not be surpassed.

SECTION III.

EXAMINATIONS

IN CONNECTION WITH

THE POPULAR LECTURES.

In the Autumn of 1868, when preparations were being made for the third session of delivery of my Course, the Rev. Mr. MURPHY suggested the expediency of testing by means of Examinations, the actual amount of knowledge imbibed and retained by those men of Lambeth who seemed so earnest in their desire for Science. I the more readily adopted this suggestion, as it afforded me an opportunity of putting to a practical test, certain innovations in the way of Elementary Examinations, which I had long been anxious to see applied to the furtherance of scientific instruction among the People. The manner in which the experiment was conducted is described as follows, in a Letter which I addressed to Mr. MURPHY at its conclusion, and which was read by him at a crowded meeting held at the Baths on the 20th February, 1869, when LORD SHAFTESBURY distributed the Prizes to the successful candidates :—

* * * * "Looking carefully through the eight Lectures into which I had endeavoured to compress the scientific facts most wanted in Daily Life,* I divided the subject matter of each Lecture into a con-

---

* A second Lecture on Chemistry has since raised the number to nine.

venient number of parts,* and prepared for each part a question so framed, that anyone answering it might fairly and freely show to what extent his intelligence and his memory served him; and of course being competent to answer all the questions would be tantamount to having in one's mind the gist of all the Course. You were kind enough to read publicly the whole of these questions at the opening of the Lecture Season, and as a copy of them was deposited at the Baths, together with a copy of each Lecture, and moreover, as each question had attached to it the maximum number of marks which an answer to it might gain, the whole scheme of studies and rewards was spread open from the beginning. Each student knew what he was recommended to learn, and was offered the convenience of learning it without the annoyance of conflicting books. Each candidate knew what questions he was liable to be asked, and what each satisfactory answer might be relatively worth in marks."

"When the Examination time came, I selected with the assistance of Mr. HUDSON, who is thoroughly conversant with matters of this kind, two of the questions on each Lecture, one easy, the other more difficult; making sixteen questions for the whole Course. Each candidate who having satisfactorily passed a preliminary Test on the 22nd of January, was admitted to the definitive one on the 5th of the present month, found on taking his assigned place at the Examination Table, a copy of those sixteen questions before him. The Candidates were informed that they were not *expected* to take up more than *one* question on each Lecture; but such was the zeal with which they set to work, that in the three hours allowed, several of them went through nearly the whole range; and this in such a manner as perfectly to justify the sanguine expectations raised

---

* There were mostly ten to twelve parts in each Lecture.

in my mind, by the remarkable earnestness with which the Lambeth Baths' audiences, young and old, had throughout listened to my Course. As for the demeanour of the Candidates at this three hours' trial of their manners as well as of their minds, I can only say that it equalled that at the preliminary Test, which is the highest praise that can be bestowed. The number of Candidates has been small, and for this reason I have abstained from inviting the co-operation of the Society of Arts; but quality has made amends. All seven have deserved Certificates of having satisfactorily passed the Examination; and besides the three to whom the highest number of marks have been awarded by the Examiners, (Mr. SELWAY and Mr. HUDSON) and who will accordingly receive this evening the three Prizes,* I have thought it right to add supplementary tokens of success to the three who stand next and not far from them in marks. Altogether, the appreciation of scientific knowledge, the power of imbibing and retaining it, and the power of expressing it in writing, evinced by the Working Men of South London, are facts most encouraging to the advocates of Industrial Instruction; whilst they reflect the highest credit on the enlightened Pastor under whose guidance this intellectual elevation has been attained."

The above successful working of my experiment has induced me not only to repeat it this season at the Baths, where my Course has been delivered for the fourth time, but also to try it at the Boatmen's Institute, Paddington, and at the Tabernacle in Golden Lane, where in pursuance of a special request, additional Prizes have been offered for Female Candidates.

I have accordingly had printed for distribution, a detailed Examination Programme, maintaining and developing the innovations which have been adverted

£3, £2, and £1.

to, but not sufficiently accounted for in my Letter to Mr. MURPHY, and concerning which a few further explanations may be desirable.

Examinations may serve two purposes. The more obvious one is to stimulate, test, and reward individual abilities and industry. The less obvious, and yet in certain cases the more important one, is to afford certain classes of the community a necessary guidance in the selection of those branches of study which may be most practically useful to them. For those who are well educated and well to do, and can afford to regard study as the mere solace of leisure hours, or as a wholesome drilling of the mind for an indefinite purpose, it may be convenient to have a set of competent Examiners ready to question them on any branch of knowledge to which they may take a fancy, without considering whether or not it may be likely ever to render them any direct and practical service. Whether this laxity of purpose should be encouraged even among that class of Candidates, is a question on which I have more than once plainly expressed my conviction to my colleagues in the council, and especially to our late lamented friend Mr. HARRY CHESTER. But be this as it may, it is certain that as regards a great part of our adult Industrial Population, such a system is quite inadmissible; for knowledge to them is a matter of necessity, not of choice or fancy. It is a matter of necessity, inasmuch as without knowledge their earnings are likely to be scanty, and their home comforts are almost sure to fail. It is not a matter in which they may exercise a free choice or obey the impulse of fancy, for they cannot make the wheel turn with Logic, or the pot boil with Latin.

Almost as much mischief is done by the perversion of sweeping axioms, as by the abuse of Scripture Texts. You may hear people say "Everyone knows best where the shoe pinches him," and as a natural corollary, they maintain that the working man knows better than anyone else where his deficiencies are, and that if you place instruction in each of the leading Sciences within his reach, he will naturally select what he wants. He may make a mistake or two at first, but he will soon right himself, for "things always find their own level;" moreover, any branch of Science he may have learned by mistake will be "sure to come to use some day;" and besides, Science of any kind affords such a capital "drilling of the mind." Such are the principles which have often prevailed, and perhaps sometimes succeeded, but which are none the less at variance with my conviction. Ignorance should not be its own guide to knowledge, and those who most require to learn, can least afford to waste their time and opportunities in learning what is not to the purpose.

Working men do not, generally speaking, require a deep and complete possession of any one Science, but a selection from the essential elements of several Sciences, which should be thoughtfully made by an experienced friend and carefully fitted to their wants. On examining these wants, we find firstly, that whatever the occupation may be from which the working man derives his earnings, he requires Science in the form of BIONOMIC KNOWLEDGE to direct the use of them; and secondly, that the great majority of occupations involve a further amount of Science in the form of TECHNICAL KNOWLEDGE. The latter will be the subject of Section 5 of this Memorandum. To give a practical illustration of the former, showing on what principles its elements should be selected, and how they should be put together, is the purpose of my Popular Course. To test and to strengthen the scientific foundation thus far provided, is the object of the special Examinations of which I am now explaining the peculiar nature. They are so devised as, firstly, to render more clear and positive the intellectual guidance afforded to Working Men, inducing an uniform distribution of their attention over the entire range of subjects spread out before them; and secondly, to bind together the different parts of the knowledge acquired, into a compact whole.

These and other points will best be elucidated by giving textually the explanatory portion of my Examination Programme. After a summary account of the Popular Lectures, it contains the following Articles:—

1. "Those essential Elements of Scientific Knowledge which it is most desirable that every adult of the Working Classes should possess, have been methodically arranged and explained in an easy and familiar manner in the foregoing Lectures."

2. "For the purpose of the Examinations, each Lecture has been divided into a convenient number of parts, large or small according to the nature of the subjects; and a question has been prepared for each part, embracing as far as possible the whole substance or gist thereof, and so framed that a person answering it might be prompted to show to what extent he had understood and retained the matter referred to."

3. "The whole of the questions, which amount to about one hundred for the nine Lectures, are annexed to the present Programme. It is from them that will be selected at each Examination the questions to be put to the Candidates."

I beg leave to remark that it is in this feature of publishing from the beginning the questions which Candidates may have to face, that lies my greatest deviation from the ordinary practice. It is true that considerable divergency has prevailed in the principles on which ordinary Examination Syllabuses have been prepared. In many instances the predominant thought in the minds of their authors, appears to have been an anxiety to baffle any attempt on the part of the Candidates to guess the probable nature of the questions that might be asked. The result of this anxiety has been to produce a hazy Syllabus, followed by questions of which the chief merit seemed to lie in their having never been thought of before, and in being such as no one could possibly expect. In other quarters we have seen Syllabuses evidently designed to convey to the Student a much clearer idea of the range of Science he was expected to master. But nowhere to my knowledge has any approach been made to the plan which I have ventured to adopt, of putting in the student's hands from the beginning, a complete list of the exact questions he may have to answer, and yet it will be found as we pursue our review of the Programme, that this plan can be rendered as safe and equitable, as it is likely to be satisfactory to those concerned.

> 4. "The questions, besides being consecutively numbered for each Lecture, have attached to them other numbers ranging from 3 to 25, which indicate their more or less comprehensive and difficult nature, and show the *maximum* number of *marks* which a candidate could obtain at the Examination by answering them in a thoroughly efficient manner."

Though the numbers stand as a rule in direct proportion to the difficulty of the questions, yet in some instances a little favour is shown to subjects of great practical importance, in order to encourage their study.

5. "It is obvious that a Candidate in order to be competent to answer thoroughly any of the questions which the Examiner might happen to select, would have to be master of the whole of the Elements of Practical Science as embodied in the above Course; but far less than this degree of efficiency will suffice for securing to the Candidates a satisfactory Certificate. The *minimum* of *marks* required for this purpose, will be determined by the Examiner."

6. "The series of questions, with maximum numbers attached, will remain the same from year to year, subject to any improvements which it may be found expedient to introduce, and to any alterations or further development of the Course. It will be freely circulated previously to and during the beginning of the delivery of the Course in each locality, together with the Syllabus of the Lectures, so that the Working Men of the District may know at once what they are wanted to learn, what questions they would be liable to be asked if they should become Candidates, and what might be the relative value of each satisfactory answer."

7. "When the present 'Popular Course' is printed, each Candidate may obtain a copy on reduced terms. In the meantime such facilities will be afforded as convenience may allow, for the inspection of a manuscript copy."

8. "Persons desirous to become Candidates in the several localities where the Lectures and Examinations may be organized, will be expected to conform to the following regulations, which will be entrusted to a Committee of supervision."

a. "Candidates must be at least 18 years old, and strictly of the Working Class. In localities where it may be found desirable to admit Female Candidates, the minimum age required for these will be 16."

b. "They must furnish for the consideration of the Committee of Supervision, the following particulars:
"Full Name, Address and Age."
"Employment, and where employed."

c. "Each Candidate approved by the Committee will receive a Card, on which he (or she) will at once inscribe his (or her) own name."

d. "As a rule, no Cards will be issued after the fourth Lecture, and no Candidate will be examined who has not attended at least six of the Lectures, and each time signed the attendance book."

9. "The Examination will be held by Mr. HUDSON, the Examiner, in the room and on the evening appointed in concert with the Committee of Supervision, such evening to be if possible within one month after the conclusion of the Course."

10. "At a Friendly Meeting of the Candidates, held sometime before the Examination, each of them will be asked to answer in writing a few very simple questions concerning the Lectures, and those only who do this satisfactorily will be admitted to the Examination."

11. "Previously to the Examination, two questions on each Lecture, one very easy, and the other more comprehensive and difficult, will be selected and marked on the Programme by Mr. TWINING in concert with Mr. HUDSON, with a view to giving each Candidate an opportunity of adapting his attempts to his abilities."

12. "On the evening appointed for the Examination, the Candidates will find before them on taking the places assigned to them at the table, besides the necessary writing materials, copies of the Programme on which the eighteen questions are duly marked. They will be informed that three hours are allowed them for preparing in presence of the Examiner, and without any assistance beyond their memory, plainly written answers to as many of these questions as they may choose; but that they are not *expected* to take up more than one question for each Lecture, and that even six good answers may earn a Prize. Minor formalities will be duly explained by the Examiner." *

13. "The Examination Papers, signed by the respective Candidates, and left by them on the table, will be brought to Twickenham for being carefully examined, and each more or less satisfactory answer will be rated at a proportionate number of marks."

14. "Each Candidate whose total number of marks reaches the amount fixed beforehand as necessary by the Examiner, will receive a Certificate."

15. "To the Candidate or Candidates who shall stand highest, a Prize or Prizes of from £1 to £3 in Money, Books, or otherwise, will be awarded, subject to the following conditions":—

a. " No Prize will be given unless three Candidates compete, when one Prize of £1 will be awarded."

b. " If the number of Candidates be from six

---

* Papers from various Examinations will be shown to any Members of our Council who may be desirous of inspecting them.

to eleven, both inclusive, two Prizes will be awarded of £2 and £1 respectively."

c. "If the number of Candidates exceed eleven, then three Prizes will be bestowed, of £3, £2, and £1 respectively."

16. "The results will be made known to the Candidates if possible within a fortnight after the Examination, and a convenient evening will be appointed in concert with the Committee of Supervision, for the distribution of the Prizes and Certificates."

The foregoing regulations are followed by the series of questions to which they refer, and which are not under nine nor over fourteen for each Lecture, making in all 103, and filling five pages quarto with double columns.*

SUMMARY OF PURPOSES.—The leading purposes of the foregoing plan may be summarized as follows:—

a. To sketch out clearly a Curriculum of Studies suited to the requirements of a given class of Students.

b. To invite competition, rendering it more satisfactory and attractive, by giving from the beginning free publicity to the whole of the questions liable to be asked at the Examinations; whereby Candidates are inspired with confidence, and secured against thoughtless, eccentric, or unfair questions.

c. To secure an even distribution of the questions to be selected, over the whole range of the Curriculum.

d. To give by the manner in which the questions are worded, and also by the rule of selecting for each Lecture one easy and one more difficult question, a

---

* Copies of the Programme, as well as specimens of Candidates' Cards, &c., will be freely supplied on application to the Secretary of the Economic Museum.

fair opportunity for Candidates to display what they may know, whether it be little or much.

e. To give permanency to the Syllabus or List of Questions, whereby increased care is probably secured in its first preparation, whilst an opportunity is afforded for improving it at each annual revision, and for keeping it up to the level of the times.

f. To lighten the task of the conscientious Examiner, for whom it is a considerable saving of trouble, to have a well digested series of questions spread out before him for his selection.

I cannot perhaps better conclude this chapter than by mentioning that in a letter lately received from M. CHARLES BULS, General Secretary of the Belgian Educational League, I have been informed in flattering terms that he intends trying Examinations on my principles as soon as he can find time to prepare a set of questions.

SECTION IV.

PRACTICAL BIONOMY

OR

THE SCIENCE OF COMMON LIFE, AS A PART OF PRIMARY EDUCATION.

I have hitherto chiefly considered my Popular Course in reference to the Adult Working Population of the Metropolis, for whose benefit it was more specially prepared; but I trust that in so doing, I have sufficiently established the desirableness of Instruction of this kind for the Industrial Community at large.

I have not filled page after page, as I might easily have done, with an enumeration of the economic blunders committed and evils incurred by Working Men and their Families in the ordinary routine of Daily Life, and this simply through the want of a good insight into the requirements of the human frame, into the nature of the resources and dangers which surround it, and into the manner of dealing with each respectively. The value of such a guiding insight for all classes, and especially for those who depend on their health and strength for their earnings, and on their own knowledge and intelligence for a good use of them, is too obvious to require lengthy arguments. Equally obvious is it that fallacious guidance would be worse than no guidance at all. Thoroughly sound as well as thoroughly practical must that knowledge be, which is to act as helmsman at all times and in all weathers, and nothing could make it such but a genuine and methodical Scientific training.

It is not any single Science, though ever so completely mastered, that could serve the purpose; not even Chemistry itself, the Queen of Sciences. If we investigate the circumstances of a Working Man's condition, his inward temperament, and the little world of facts, influences and contingencies with which he is more immediately surrounded, we find here Chemistry required to help him in his daily struggle, there Physics equally indispensable; then, more frequently still, an amalgamation of the two is necessary, whilst everywhere Physiology must be at his elbow, and occasionally Natural History has a word to say.

At first one is startled at the mass of information apparently required; but on close examination one finds that only the fundamental facts and simplest principles of these various Sciences are absolutely necessary, or at all events that with these a large majority of the obstacles which beset the Working Man may be overcome, provided he possesses naturally, or has acquired through training, the faculty, and what is more the habit and the inclination, of thinking logically and quickly, of putting two and two together, and his shoulder to the wheel. One finds that the Physics he requires do not necessarily involve any difficult mathematical problems, that his modicum of Chemistry leaves untouched considerably more than one half of the elementary bodies, and a vastly greater proportion of the compound ones, that the long list of inorganic serials and substitutionals, may be ignored, that for him a starch is not "an oxygen ether, or anhydride of a polyglucosic alcohol of a high order,"* and that he can even learn what he most wants, without symbols, equivalents or algebraical notations.

* See Fownes' Chemistry, 10th Edition, page 684.

It is on investigations and conclusions like these, that is based the range of scientific knowledge, elementary and applied, which I have described under the name of Practical Bionomy, or the Science of Common Life ; the applied knowledge embracing in natural order the various departments of Household and Health Economy, and the elementary knowledge embracing the indispensable quantum of preliminary or preparatory Science, indicated by the present nine Lectures of my Popular Course. I do not particularly wish to uphold the exact circumscription of Subjects which I have adopted, but I trust it will be found near enough the mark to afford a substantial base of discussion, and a convenient starting point for kindred endeavours. Let us then assume that something like this kind and amount of Science is what our Working Classes want to make ends meet, and to help them fight the battle of life. It is evident that the earlier they get it, and the more indelibly it is impressed on their minds, the better. Here then arises the question, is there not a way of propagating it, infinitely more comprehensive in its action, and more secure in its results, than the offer of Lectures to an adult population of whom comparatively few can avail themselves of the opportunity, whilst of these again, a small proportion only can be expected to derive much permanent benefit ? Might not each rising generation receive in due time the essential principles of scientific guidance as a part of its School Education ? *

The first doubt that presents itself to our minds, is whether it be possible to infuse Science into those of children ; but this doubt falls to the ground when we

---

* In order to simplify the question before us, I avoid discussing any amount of Science which the children of the People should be taught beyond that now proposed on simply utilitarian grounds. It is evident that even the son of the labourer should possess those primordial notions concerning the Earth, Sun, Moon and Stars, without which he would dishonour the human intellect with which he is endowed ; but these matters may be considered as belonging to the same class of studies as the rudiments of Geography, History, &c.

see that scientific knowledge of a thoroughly earnest and methodical nature has, through the mere circumstance of its being well illustrated and couched in familiar and cheerful language, been rendered intelligible and palatable to audiences largely recruited from the most illiterate ranks of society; from those ranks where so many adult minds are at the present time, as far as culture is concerned, actually in a state of infancy. The fact is that my Popular Course, as I candidly announce in Lecture I, begins with the very A.B.C. of scientific language, and the account given of the Physical properties of Bodies, is in many instances as rudimentary as that taught in Schools under the name of "KNOWLEDGE OF COMMON THINGS." As the Course progresses, I take advantage of the information which the audience may be assumed to have acquired, and of their increased aptitude for imbibing knowledge, and gradually raise the tone of the instruction. In doing so, I cannot help feeling continually that the narrow limits necessarily prescribed to a Course of this description, compel me to hurry upward and onward faster than many of my industrial friends, unaided as yet by a Text Book, can possibly follow me. Nor have I any means of varying the scale of difficulties according to the scale of abilities, and I often envy the School-master who could at least *sort* the little minds he would have to deal with, and let his instruction grow *pari passu* with their growth.

Before proceeding further in this direction, it may be well to attach a clear meaning to particular terms. A fruitful source of misunderstanding and of interminable discussions, is the want of some preliminary agreement as to the meaning of certain words. Two Educationalists who are actually of the same mind, may be seen at logger-heads, because the one means by Religion what our Rev. Friend Mr. ROGERS has called GENERAL RELIGION, whilst the other thinks that DOGMATICAL or SPECIAL RELIGION is meant. And so likewise one who after the fashion prevalent in some parts of the continent, would have every boy stay at school till 14, reckons all as PRIMARY instruction that is taught up to that age, and quarrels with his friend who would teach exactly the same things at the same ages, but makes in his mind two stages of the educational journey, and reckons as PRIMARY what comes under 11 or 12, and as SECONDARY what follows up to 14. Now I do not wish to pin my scheme of early and progressive science-teaching in the Schools of the People to either of these modes of reckoning; nor indeed to any of the plans for limiting or extending the school years, for employing them on the whole or the half-time system, or for rendering the instruction either denominational or unsectarian. I equally set aside for the present the question of the difference to be made between Boys and Girls, and will simply assume by way of hypothesis, and to facilitate explanations, that all boys of the Industrial Classes will go to school from 6 or 7 to 12 years of age, which period I will call PRIMARY ; and that those intended to become Artisans or Working Tradesmen, or whose parents can well afford it, will go to school for a further period from 12 to 14; which period I will name SECONDARY. Adapting my proposals to this plan, (and they might equally well be adapted to many others,) I venture the following propositions:—

1. Many of the visible and tangible features or properties of COMMON OBJECTS, which children at an early age are found capable of noticing and retaining, might advantageously be amalgamated with illustrations of simple scientific facts, and worked into a kind of visual entertainment or show, in small progressive parts, which would be all the better remembered for being mixed up with anecdote and fun.

2. These properties and facts might still better be impressed, whenever opportunities presented themselves or could be created, by their being brought in to meet a want, solve a difficulty, or confer an advantage.

3. After a time the properties and facts, which were before matters of amusement or opportunity, might be renewed in connected sequence, and separated into groups belonging to distinct Sciences, of which the names would now be introduced.

4. Going a third time over the ground, the teacher would add to the elementary knowledge thus consolidated, the applied knowledge to which it was designed to serve as a foundation. Thus he would make his instruction approximate by degrees in range and character, to what I have described as PRACTICAL BIONOMY, and before the expiration of the primary period, he might succeed in storing the minds of his pupils, if not with a very notable, yet with a very serviceable amount of Bionomic Knowledge. This it would be his duty to select with discrimination and forethought; carefully considering what, in view of the probable condition and occupations of the boys, would be most likely to render them essential service, what might be made lively and interesting, and have a chance of being understood and retained, and if possible further developed, and what would be consistent with the rules and resources of the School. What he should however prize and foster more than any amount of facts and precepts stored up in the memory, would be a ready ability to bring them to bear on the right purpose at the right time. It is through common sense and presence of mind, that mere Scientific Knowledge becomes the Science of Common Life or Practical Bionomy; ever thoughtful, vigilant and active, ever ready to secure a legitimate benefit or to avert an injury; having a word of advice for every difficulty, and a word of comfort for every mishap; a trusty guide at all times, and a true friend in the hour of need.

5. Assuming that boys intended to become Artisans or Working Tradesmen would, as said above, have the benefit of a secondary schooling from 12 to 14, we might hope that this very capable period would allow

of such a further development of the scientific knowledge previously acquired, as to make it serve, not merely for *Bionomic,* but also for *Technical* purposes.

6. In reference to the educational materials required for such a course of instruction, I may remark that as regards articles of Domestic Economy, those chiefly studied should be "common things." As regards the required Books, Diagrams and Apparatus, they will form the subject of future portions of this Memorandum. I may however mention here, that a considerable portion of the illustrations used in the physical part of my Popular Course, are such as almost every person able to handle a Joiner's tools could prepare with inexpensive materials, and that in many other instances I use children's toys; a fact suggestive of the ease with which one might in a thousand cases, make Science find its way to little heads through cheerful hearts.

7. Summing up the data in hand, viz : *a,* the evidence supplied by the audiences at my Lectures, in most cases utterly destitute of all previous scientific training, and in many cases prejudiced against Science by previous samples of the wrong sort, and yet withal so studious and appreciative that they evidently only want to be able to attend regularly a series of colloquial classes of the same character, to become possessors of a considerable fund of practical knowledge ; *b,* the evidence supplied by many existing Schools, and especially those conducted on the Pestalozzian system, as to the capability of children of learning things demonstrated visibly and tangibly, vastly better than those which are matters of thought or memory, and especially things of which they see the purpose or enjoy the zest, far better than those which they only know to be deserving of attention because they are told so; considering these and many other data of a similar character, the enumeration of which would be tedious, I come to the conclusion that as far as the children themselves are concerned, the Primary Education of the People might perfectly

be made to include the most essential portions of Practical Bionomy.

But here two questions stop the way. The first is whether time could be found for introducing so important an addition to the old primary routine. I will candidly confess that if I had thought that the old routine was intended to be persevered in, I should scarcely have written these pages; but I have written them because, on the contrary, I see men's minds expanding with a conviction that children's minds are also capable of expansion; because I see common sense taking the place of prejudice, and that kind of knowledge coming into favour which may best help the Working Man to help himself, and because at the same time I see that, thanks to the labours of our enlightened educationalists, the art of teaching the masses is tending towards a change, almost as great as that which manufacturing Art formerly underwent through the introduction of machinery.

Much more ominous is to my mind the second question; viz, whether Teachers can be found to carry out a scheme evidently involving a more comprehensive knowledge of Science than it is at present the good fortune of many to possess, and also requiring a rather unusually thoughtful ponderation as to what practical knowledge the rising generation of Workmen in general, and of the Workmen of a given locality in particular, are likely to stand in need of; to say nothing of the troublesome responsibility of adapting instruction to circumstances, instead of carrying it on according to the "rule-of-thumb."

I cannot deny that my scheme, (in which I will now include the education of Girls, previously omitted for the convenience of argument,) calls for a specially trained host of Teachers, Male and Female, not conspicuous for their proficiency in one Science and their ignorance of the rest, but possessing a well selected and well arranged assortment of scientific knowledge, elementary and applied, embracing the whole of the normal requirements of Common Life.—Now of late years Science in general has, for reasons too well known to require mention, been at a discount among primary Teachers; and as for this composite kind of Science in particular, it has been quite out of the question. Students at the Training Schools might gain something by learning Chemistry, Physics, or other branches of Science, but each branch would be taken upon its own merits, without any acknowledged connection with the others, or with the practical requirements of Daily Life. On the other hand there have been Examinations in Domestic Economy, but they did not rest on a special scientific foundation, and the consequence was what might have been expected: a knowledge of facts unsupported by a knowledge of principles, which the least shaking would upset, and rules without rationale, which would only fit where everything was cut according to a given pattern.

In order to remedy these deficiencies, two things would be required: Firstly, to supply present or future Teachers with opportunities and incentives for learning what it is desired that they should impart; Secondly, to ascertain by suitable examinations, whether through these or any other means they have arrived at the kind and degree of competency required. I have thought I should do most good by taking the latter desideratum first; trusting that if I could induce by the offer of Prizes the coming forward of a certain number of Candidates possessing the qualifications desired, the feasibility and usefulness of my plan might be demonstrated, interest and emulation might be raised, and an acknowledged demand for special training would soon be met by a regular supply. Accordingly I prepared in 1867, for the consideration of the Science and Art Department, a scheme of Examinations, of which the nature may be seen by the following letter and enclosure, addressed to our worthy friend Mr. COLE on 19th June of that year.

My dear Sir,

"Allow me to submit for the consideration of their Lordships the accompanying Memorandum, giving the details of a proposal which I beg leave to make for placing at their disposal a sum of £70, to be distributed in Prizes to Schoolmasters, Schoolmistresses, and Teachers, at Competitive Examinations, which would be held in the Spring of 1868, for a comprehensive and scientific knowledge of the subjects which have the most direct bearing on the health and comfort of the Working Classes. Of these subjects my Memorandum includes a detailed Syllabus."

Believe me, &c., &c.,

H. Cole, Esq., C.B.   T. Twining.

The Memorandum adverted to contains the following explanations and remarks:—

"The £70 could be distributed in the following Prizes:—

Males—£20, £10, £5.   Females—£20, £10, £5."

"The Candidates should be persons engaged as Masters, Mistresses, or Teachers at National, Parochial or other similar Schools for the Children of the Working Classes, or whose intention of devoting themselves to this line of scholastic duties is evidenced by their being or having been pupils at Normal Schools. They should not be above 25 years of age, nor in the receipt of a higher salary than £100 per annum in net cash."

"The Examiner or Examiners would be appointed by Mr. Twining, subject to the approval of the Department of Science and Art."

"The title of Practical Bionomy is proposed as blending, under a brief denomination, the scientific principles and practical rules which should guide us in Daily Life, and which are generally included under the various names of Hygiene and Domestic Economy, of

Domestic and Sanitary Economy or of Household Economy and the Laws of Health. Should the title of 'Practical Bionomy' be objected to, I could substitute 'Science of Common Life.'"

"From the fact that the attainments to be tested are to embrace the practical applications of several Sciences, two considerations result :—

1. "It is necessary that they be based on a sound knowledge of the essential elements of those Sciences, and on a sufficient acquaintance with their phraseology."

2. "It is necessary, as it would be in the case of agriculture, or of any compound Science made up of selections from other Sciences, to establish at the outset a circumscription which may appear to be as little as possible arbitrary, and as much as possible founded on experience and common sense. This is doubly necessary in a department of studies of comparatively recent development."

"The following Syllabus (in 12 Sections) has been prepared in conformity with these considerations. The first four Sections are intended to show to what extent the Elements of Physics, Chemistry and Physiology, with the general outlines of Natural History, constitute an indispensable foundation for the study of the requirements, resources and processes of Daily Life, and consequently to what extent a knowledge of them may legitimately be demanded of those who aspire to become the Teachers of the People; the more so as it has been practically proved that Science may thus far be rendered easy and attractive, even to the uneducated, without ceasing to be strictly methodical. The remainder of the Syllabus (Sections V to XII) shows what departments of Household and Health Economy are considered as deserving to become a standard element in the school education of the children of the People, and in the Evening Class Instruction of Adults, and of which therefore a competent knowledge should be possessed by their Instructors."

"The Examination Papers, for the working of which three hours would be allowed, would contain questions on, say three of the subjects indicated in each of the twelve Sections of the Syllabus, and the Candidates would be expected to answer at least one question in each Section."

"No question would be asked which is not directly indicated or naturally suggested by the Syllabus."

"The questions would be comprehensively framed with a view to give all Candidates full and fair scope for making manifest their respective amounts of knowledge in each department. A sufficient knowledge of many essential subjects, would be preferred to a deep knowledge of a few."

"It is assumed that the Syllabus contains no subject of which it would not be desirable that all Candidates should possess at least general notions. Nevertheless some of the subjects claim more specially the attention of Male Candidates, others that of Female Candidates. They are indicated by (M) or (F) prefixed to the Paragraphs which contain them. An appropriate distinction would be made in preparing the Lists of Questions."

I omit the Syllabus which was appended to the foregoing communication. Its first four Sections were but a modification of the Syllabus of my present Popular Course, *(See Appendix No 2,)* whilst the other eight Sections compassed in a similar way the contents of its intended sequel.* If my Course had been completed and printed, it might to a certain extent have served the purpose of a Text Book; but I felt that something of a more special character was wanted to form the basis of the proposed Examinations, and I began preparing a book in a catechetical form, intended not only to supply the knowledge which I considered that the teachers ought to be able to communicate, but also to show them in detail the manner in which that

---

* They accordingly corresponded in some measure to the range of subjects illustrated by my Economic Museum. (*See Appendix No.* 1.)

knowledge might, as I thought, be best communicated to youthful minds, according to the principles of selection, adaptation and progression above explained. A considerable portion of this Catechism is in manuscript.

My idea was that the Department of Science and Art should undertake the publication of my scheme, and the office work connected with the proposed Examinations, but that I should provide the Fees for first-rate Examiners. I reckoned that these might raise the outlay from £70 to £100, which expense I was willing to incur for three successive years, hoping that by that time the utility of the plan would become sufficiently apparent, to induce its being taken up by our educational authorities on their own account. For a time there seemed to be a likelihood that my proposal would be entertained; but in December, 1867, I was informed that the Science and Art Department could not take charge of conducting the Examinations in question on my behalf, unless my Syllabus were entirely surrendered to their Lordships. To this arrangement I declined to accede, fearing it might lead to the omission or alteration of the particular principles according to which I was anxious that the selection and arrangement of subjects should be made, and the examinations themselves conducted. Since then the successful working of these principles in a small way in my Popular Examinations, has increased my desire that some opportunity may present itself for testing their practical value on a suitable scale, and my willingness to incur any necessary expense. *

* I should be happy to show to any friends interested in these matters, the Syllabus referred to and other papers connected with this special movement, including a comprehensive classified list of references to the most eligible books in the various departments of Domestic and Sanitary Economy. The selection of these books was made so as best to suit the purpose and come within the means of popular teachers.

SECTION V.

SCIENCE AS A PART OF TECHNICAL INSTRUCTION.

I have endeavoured to show in the preceding sections: firstly, that Working Men are perfectly accessible to scientific knowledge, provided it be of a practical character, and offered to them in an easy and entertaining form; and secondly, that similar knowledge might in all probability be gradually instilled in an analogous form into the minds of the Children of the People, as a part of their School Education, by Teachers duly trained for the purpose. The scientific knowledge referred to is that which I have found convenient to call Practical Bionomy, being Common Sense improved with scientific acumen, and made into a Code for regulating the practical concerns of Daily Life, to the obvious benefit of the intellectual and spiritual ones. Such guidance no social class should be without, but least of all the Working Man; not only because he cannot afford to make mistakes with his health or his money, but also because the amount of Science, especially Physics and Chemistry, which forms the ground-work of Practical Bionomy, and the habit of turning scientific knowledge to practical purposes, may prove a valuable furtherance to him in mastering his Trade. There are many occupations for which this small amount of elementary and practical Science is abundantly sufficient, but there are many others which involve a considerable amount of scientific knowledge, whilst others again recognize Art rather than Science as their leading star. I will for the present confine my remarks to the scientific element in Technical Instruction, showing what I consider to be the chief desiderata, and in what manner I am endeavouring to supply one of them.

In the same manner that we have seen Practical Bionomy dividing itself naturally into two parts, the one more particularly devoted to elementary science, the other to the practical matters on which that science is to be brought to bear, so likewise do we find that the instruction required for Trades involving scientific knowledge, separates itself into an elementary part in which the scientific teaching may be more or less the same for two or more Trades, and a purely special or technical part.

Nothing is more useful for acquiring a clear notion of the parallelisms and divergencies of the various Trades, and of the extent to which young men in training for them may be taught together or must be taught separately, than to inscribe a number of Trades more or less scientific, artistic, or manual, in the first column of a Synoptical Table, having another column for each branch of knowledge or attainment. In lately looking through some papers written about twenty years ago, when I was vainly striving with a few friends to raise the cry of scientific progress, I discovered among them the sketch of such a Synopsis. It was never completed, but its framework remained in my mind, and has often rendered me great service in endeavouring to devise the means, by which the greatest amount of sound Technical Instruction, might be imparted with the smallest expenditure of Teaching Power.

One of the facts thus strikingly demonstrated, was the great predominancy of Mechanical Physics, Chemical Physics and Chemistry among the scientific ingredients of Technical Instruction. It was evident that Chemistry in particular ministers to a vast number of Trades, of which some delight more especially in its metallurgical departments, whilst others less inclined that way, embrace in other directions a very considerable range of Chemical Science. I saw that among the latter category of Trades, that of the Dyer is particularly comprehensive in its grasp, and that if I could bring out a Course of Lessons calculated to satisfy the whole of its ordinary chemical requirements, I should include those of most other Chemical (not metallurgical) Trades. I felt that this would be a substantial ground of study common to the whole category, and that I should only have to add a certain quantum of special elementary and technical teaching for each trade, in order to give a practical example of the manner in which I conceived that industrial studies should be carried on.

It is right to explain that the Chemistry to which I here allude, is the practical and comparatively easy kind required by the Working Dyer, and not the higher kind of Chemistry that should be possessed by the Foreman of a large Dyeing Establishment. It is not necessary to discuss at present the various ways in which Chemistry should be studied, according to the previous preparation of the student, and the purpose he has in view. It is obvious that in dealing with Working Men and Lads brought up under the old educational system, and who are to take up Chemistry with minds totally unprepared, one must absolutely set aside the *beau idéal* of the perfect Science, and be only too glad if, by softening down certain difficulties and avoiding others, one can succeed in introducing into these untutored minds, a tolerably complete and methodical sequence of sound facts and principles, calculated to render practical service. Such is the nature of the ELEMENTS OF INDUSTRIAL CHEMISTRY comprised in the 24 Lessons the preparing of which has been my chief occupation for the last two years. Eight are devoted to Inorganic, and sixteen to Organic Chemistry. I shall perhaps add a 25th Lesson, for the purpose of giving to those who might be induced to pursue further their chemical studies, a few explanations tending to render less puzzling the use of the various Handbooks written according to different systems.

It is right I should mention the following as circumstances that have mainly contributed to induce me to engage in this rather arduous undertaking. Firstly, the love for Chemistry imbibed at an early age at Paris under Professor Orfila, and followed up whenever opportunities allowed; secondly, the loss of voice which during these last two years has favoured the occupation of writing by checking other enjoyments; and thirdly, the valuable assistance on which I could rely on the part of Mr. HUDSON, who acts as the Chemical Superintendent of my Museum. His critical and conscientious researches in scientific literature, have been exceedingly useful to me in collating the best authorities on disputed points, and in collecting scraps of information required for my particular purpose, and not found in most of the Treatises on Chemistry. I also owe to his exertions in the laboratory, the neat and complete sets of Illustrations, which have been packed by the Curator, Mr. FREEMAN, in appropriate boxes for circulation, and amounting in the aggregate to more than one thousand articles.*

In the style in which these Lessons are written and got up, they resemble the Popular Lectures, but they incline less towards recreation and more towards earnest study. They are shorter, being calculated to last about an hour if read by the same person who demonstrates, and considerably less in the binary mode of delivery. Thus time will be allowed for the students to question the teacher on any points they may not have understood, and for the teacher to precede each discourse by examining the students on the previous one. In short the

---

* Of the 24 sets the greater number are ready for use, and I should be happy to submit them to the inspection of the Members of the Council, or to display them at next year's Exhibition of Educational Materials.

whole is intended to have the character of a regular and earnest series of Class Lessons.

Let us now consider this chemical instruction in connection with the other portions of instruction, elementary and technical, which as I have said above, will be respectively required by the various Trades. We have seen that the Dyer is at the head of the Chemical Trades, and we may take his requirements as representing those of other members of the group.

The Dyer, like everybody else, should possess that scientific knowledge, elementary and applied, which I have assumed to be the best secular key to health and comfort. I hope that in ten or twenty years hence, there will scarcely be a Tradesman or Artisan who has not acquired that knowledge to some extent in his Primary Education, and further developed it, with additions in a technical direction, during the important school period from 12 to 14; but in the meantime we must take Tradesmen and their Apprentices as the old system has made them, and I will select by way of specimen an ordinary Dyer's Apprentice, desirous of securing success in the practice of his Trade by mastering its rationale. I first take the liberty of recommending to him my own Course of "Science made Easy," not knowing any introduction to Science on which more pains have been bestowed in order to conduct the beginner from the very bottom of the hill to a fair distance up it, without letting him feel that it is steep, or making him push his way through brambles, or causing him to tread on any unsafe ground. Our young Dyer will find, what I have already asserted, that the elements of Science required for Daily Life go a considerable part of the way towards meeting the requirements of Technical Industry, and he will do well to store in a handy corner of his memory, the physical laws illustrated in the first four Lectures of my Popular Course. Lecture IV in particular will teach him much that is to his purpose, but not enough, especially as regards Light. I have taken notice of the deficiency and am consequently preparing supplementary physical Lessons for the use of Dyers, giving a tolerably full account of primary, secondary, and tertiary colours, complementary colours, harmony and contrast, and other matters by the knowledge of which the Dyer may be rendered an intelligent coadjutor, rather than a passive instrument of manufacturing industry on the one hand, and of fashion on the other.

I have alluded, in speaking of the manner of teaching Science to children, to the advantage of going two or three times over the same ground, and of adopting each time a further development of the subject proportionate to the mental development of the pupils. The benefits of this plan, of which I hope to say more elsewhere, are by no means confined to children, and the young Dyer will not have any cause to regret having taken the rudiments of Chemistry in two Lectures (V and VI of my Popular Course) as a prelude to my ELEMENTS OF INDUSTRIAL CHEMISTRY in 24 Lessons. These latter I must ask him to go through in good earnest, as they are intended so far to complete the *first* or *elementary* part of his Industrial Knowledge, that he may be competent to study intelligently and scientifically the *second* part, consisting of applied and more strictly *Technical* Knowledge.

In order to secure for this important part of the Dyer's studies, reliable information which might at the same time serve to exemplify in a practical way my proposed plan of instruction, I have had prepared by a person well acquainted with Chemistry, materials for a series of Lessons on the Art of Dyeing, embracing the following subjects :—Historical Sketch. Textile Materials. The Dye House. Dyeing Materials. Dyeing Processes. Calico Printing. Cleaning and Scouring. Special Pathology of the Dyeing Trade. These materials have already been submitted to a competent master Dyer, and what remains to be done is simply to arrange them in Lessons, similar to, and in strict accordance with, the chemical series.

I have likewise in hand, draft sets of Technical Lessons by the same writer, for the Tanner and Currier, and the Plumber and Glazier. The Popular Course as a first scientific foundation, and the 24-Lesson Course of Industrial Chemistry as a more earnest insight into that most important Science, will serve for these Trades in common, but each will have supplementary Lessons on any branch of elementary Science that may be required, and each will conclude with its own special Course of Technical Lessons.

If we pass to the Mechanical Trades, we find that there is everywhere scope for the same utilitarian principles of forethought, ponderation, and amalgamation: FORETHOUGHT, taking note of the items of knowledge that each Trade is likely to require normally or incidently, PONDERATION, weighing these items against each other, that only the most indispensable may be recommended to those who cannot possibly be expected to learn all, and AMALGAMATION, uniting the studies, and consequently improving the resources for study, of the various Trades, wherever they tread a common path. Thus the Carpenter and Joiner may find that the rudiments of Chemistry given in my Popular Course, are sufficient for enabling him to understand any chemical expressions that may occur in his technical lessons ; but on the other hand he must have a good insight into the manner of growth of Exogens, and should know something of the nature and habitat of the chief Timber Trees, all which knowledge would be specially supplied. Then again his Mechanical Knowledge should be well developed, he should be well up in certain branches of Arithmetic, and possess a few notions of Geometry ; and last not least, his hand, his eye and his judgment, should be duly trained in the appropriate departments of the Arts of Design.

It would be tedious to follow out with words, the

exigencies of the various handicrafts, apportioning to each its proper kind and quantity of Science and Art, and taking note of the parallelisms which might enable a certain number of them to draw knowledge from the same source, and of the divergencies which would involve separate tuition. The best, not to say the only way to arrive at a clear conception of those exigencies, is as said before, to classify and tabulate them, and this is a labour that will be found well worth the while of those who aspire to establish with economy of means and certainty of results, a national system of Technical Instruction. But at the same time it must be borne in mind that this is only one of many points on which a careful investigation is indispensable, for safely legislating with a view to improve the intellectual, technical, and social status of our industrial population. Matters like these demand the concurrent efforts of many workers, and I intend to devote the next Section of this Memorandum to an enumeration of points, on which I would suggest that communications should be addressed to the Society of Arts by all who, through special circumstances, are placed in a position to contribute reliable information and advice.

SECTION VI.

THE EDUCATIONAL WANTS OF OUR ARTISANS.

The great question of Technical Instruction has assumed a new aspect since the Paris Universal Exhibition of 1867. The evidence as to the resources of foreign Artisans, and as to the wants of our own, which was brought to light on that occasion, and which was so forcibly expressed in the Reports of the Workmen deputed by our Society, supplied the "missing link" which was required for bringing home the arguments of twenty years to an actual bearing on the public mind. Up to that time the best friends of Industrial advancement, those who knew and appreciated the exertions of foreign nations, had been striving in vain to draw attention to the rapid progress made in the race of technical improvement by our continental rivals; a progress which was just what might have been expected from the fact that they had been pulling hard whilst we had been pulling easy, and that they had been wide awake to the fertilizing influence of Science and Art on Manual Industry, whilst we had been dozing under the drowsy influence of excessive self-reliance.* The sudden change which supervened was also what might have been expected. To torpid apathy succeeded a feverish excite-

---

* People were too ignorant of Science even to feel their ignorance. This lamentable fact is forcibly brought into relief by Mr. BARTLEY's interesting Paper on Science Schools in the *Society of Arts' Journal*, of the 14th January, 1870, where he describes the apathy with which the offers of the Science and Art Department were met by local Institutions.

ment. Everyone called out for Science whether or not he understood what Science really was; everyone declared that to provide technical instruction was "the thing to do," and proclaimed his patriotic readiness to lend a helping hand. We have already technical Institutions, of a more or less genuine character, springing up in all directions; and the task of the Technical Economist is now not so much to agitate, as to take advantage of the prevailing agitation; not so much to multiply new schemes as first of all to examine impartially, and if possible to adopt and utilize existing ones. Of course, among the host of originators who are striving to convince themselves and others, that what suits their own purpose is just what the nation wants, many are egregiously in error. Some of the institutions proposed, would be scarcely worth the ground they would occupy; others that might be provisionally useful, would afterwards only be in the way; and here and there we see attempts to graft technical scions on existing educational stocks that are quite of the wrong sort for receiving them; still there are undoubtedly many notions afloat, and actual experiments in progress, that are susceptible of being turned to excellent account, provided they be taken hold of whilst yet in a plastic state, and adapted to the place they should occupy in a National System of Technical Instruction. To collect the scattered materials available for such a system, to submit them to a preliminary elaboration, and to elicit further originations where gaps appear, is a labour for the accomplishment of which the Society of Arts is pre-eminently qualified; but the responsibility of devising a definitive scheme of Technical Education, must rest with the central administrative authority which will have to superintend its realization, as a part of the general Educational System of the country.

CENTRALIZATION stands in bad repute with many people who would do well to sift the causes of their aversion. A frequent and to a certain extent a legitimate cause lies in the evil results which have accrued in some countries, where Centralization has assumed a restrictive form, hampering the independence of local action, and thus forming an obstacle to progress. But the Centralization which I should wish to see established in the Educational System of this country, would simply claim from those engaged in local efforts, a voluntary compliance with regulations established for national benefit, and such compliance, besides resting on patriotic motives, would be partly requited by direct and tangible advantages, to which I shall presently have occasion to advert.

Another objection to Centralization arises from the extraordinary diffidence which prevails in many quarters as to the manner in which things are managed in Government Offices. In fact it has become a rule with many people, that in this country everything that can, should be done by independent exertions, without any interference on the part of Government, even in the form of assistance, and I have heard in that respect a very intelligent economist quote the phrase, "Timeo Danaos, et dona ferentes." Various causes have contributed to this prejudice. Our administrative machinery is not so logical or practical as might be wished. In many cases the anomalous distribution of attributes among various ministerial departments, and the peculiar organization of these, are the result of circumstances which have ceased to exist, and are not suited to the enlightened and progressive spirit of the present day.* Retrenchment has been making itself keenly felt; and yet from time to time the press holds up to ridicule almost incredible instances of prodigal red

* LIEUT. COL. STRANGE, F.R.S., in his admirable paper on the proposed enquiry by a Royal Commission into the relation of the State to Science, remarks as follows:—"Almost every Department of the State has charge of some scientific institution—the Admiralty of one, the War-Office of another, the Board of Trade of a third, and so on, a dispersion absolutely prohibitive of harmonious system, of progressive improvement, of efficient superintendence, of economy in expenditure, and of definite responsibility."

tapeism; sheets of foolscap filled with signatures for authenticating the most trifling account, and large sums spent in the disbursement of small ones. Then again one has seen certain expensive notions pervading from top to bottom the bureaucratic atmosphere; not with any consciousness of being wasteful, but rather through a wish to do honour to the national purse. If we add to all these real or imaginary grievances, the prevailing idea that Government appointments have from time immemorial, been regulated rather by political or social motives than by a genuine desire to put the right man in the right place, we can scarcely feel surprised that a large portion of the public, without suspecting Government Officers of any wilful malversation, should hitherto have shrunk from entrusting to them any piece of management that could be managed in any other way.

Fortunately, there is in a free country no reason why abuses should exist for ever. It has been lately said, "The People's Parliament will see that the People's Government does its duty." Now I hope that the People's Government will be too proud of its name to require any incentive to do its duty, beyond the desire to merit and obtain from all classes of the People, a full measure of confidence and support, so that it may be enabled to organize, develop and administer for the general good, those departments of our social system, which cannot possibly be placed on a satisfactory footing without CENTRALIZATION.* Pre-eminent among them stands NATIONAL EDUCATION, which to attain the desired development, must evidently be centred in a powerful authority firmly constituted on a broad foundation, and culminating in a distinct personal responsibility

---

*At no period of our history has there been so great a readiness to place administrative power in the hands of the Government. Public opinion acts now so energetically and effectually in the legislature, that the old jealousy of Government interference has been almost entirely dispelled. See LIEUT. COL. STRANGE's Paper, before referred to.

I here take National Education in that broad sense in which it should be compassed in determining the attributes of the MINISTER OF PUBLIC INSTRUCTION, whose appointment was so unanimously advocated by the eminent educationalists assembled at the Society of Arts, on the 7th of February last. It is clear that his responsible supervision should include the several stages of intellectual culture corresponding to the several social levels, as well as the cross divisions resulting from diversity of Profession in some of the strata, and of Handicraft in others, and that it should equally be entrusted to his special care, to stimulate general proficiency, and to recompense exceptional merit, in the theory and practice of Science, the Fine Arts, and Music. As regards departmental organization for a division of labour according to the best interests of the public service, it should be devised by those who are privileged to consider it from a practical point of view, and I shall endeavour to steer clear of this question in the following brief analysis of a particular category of educational wants; viz, those of our Artisans. It is intended, as stated at the conclusion of the previous Section, to elicit on certain points reliable information and advice, which will doubtless find a ready means of publicity in our Society's Journal, provided they be the fruits of special experience and mature reflection, presented in a condensed form.

A.—Primary Education.

One of the first conclusions that were come to when the public mind abjured its apathy and began discussing the subject of Technical Instruction, was that little headway could be made without first improving our system of Primary Education, and this latter branch of the subject has for various reasons excited so lively an interest, that it seems at the present time almost to monopolize public attention. I have not for so many years taken a lively interest in educational matters generally, both in this country and abroad, without acquiring rather positive opinions on many of the points which are now under discussion, but I see so much controversial acrimony mixed up in the debates that are going on both orally and in print, that I feel averse to joining in them. My stand-point will be simply this, that whether the Union or the League should prevail, or whether a medium course should be adopted, I hope that the future system of popular education will be so devised and managed, as to include in its primary curriculum, all the essential elements of the Artisan's welfare, particular attention being paid to such points as the following :—

Firstly.—Moral Training. It should incorporate industrious habits and good conduct in the working man's existence, as a part of his nature, and it should prepare him for mastering the difficulties of daily life, by making him master of his own mind. We are told by our worthy colleague Mr. Bartley, who in the course of his highly praiseworthy educational investigations, has favoured us with a valuable paper on the Birkbeck Schools, that in these, and notably in the one at Peckham, a moral tone prevails "considerably above that met with in most other Schools."* As these schools are strictly unsectarian, their mode of dealing with the moral element might be deserving of particular attention if the views of the League should prevail. That its partisans are by no means unaware of the importance of this point, is clear from the following passage in Mr. Bright's late speech at Birmingham:—"In every school love of truth, love of virtue, the love of God, and the fear of offending Him should be taught."

SECONDLY. — INSTRUCTION IN SCIENTIFIC AND PRACTICAL KNOWLEDGE. It has been seen by what I have said in the preceding Sections, that the elementary and scientific instruction which I recommend as susceptible of being acquired without much difficulty by totally untrained adults, and with ease by well trained children, does not consist of any one whole science, but of a selection of the most simple and at the same time most useful facts and principles of Physics, Chemistry, and Physiology, with outlines of Natural History; these various elements being arranged as a connected series, and brought to bear practically on the wants and resources of Daily Life. The greatest possible care and forethought should of course be bestowed on the choice and adaptation of this PRIMARY SCIENCE, and indeed it should be so easy and yet so useful, that future generations may wonder how people managed to get on so long without it. The first rudiments might be insinuated into the youthful mind at an early age, by exciting little curiosities which Science might be made to satisfy, and by raising little difficulties

* See *Society of Arts Journal*, 17th Dec. 1869, page 97.

which science might be made to overcome; and this plan should as far as possible prevail throughout; every available means of rendering the instruction visible and tangible, impressive and entertaining, being pressed into the service, so as to make scientific lessons a boon and a treat, never a bore. In going two or three times over the same ground the range of ideas would naturally expand each time, so that by the end of the school years, it might embrace in all essential particulars the scientific knowledge, both elementary and applied, which I have explained in Section 4; due regard being of course had to the sex and proposed career of the scholars, as well as to the limits of time and means. I am aware that to many of my friends, it may appear somewhat Utopian to speak of introducing science, even of the easiest kind, in the education of the people, whilst there are yet millions who can neither read nor write; but with such documentary evidence before us as that which, thanks to the enlightened exertions of Mr. CHADWICK and others of our colleagues, has lately appeared in the columns of our Journal, and when especially our Rev. friend Mr. ROGERS comes to us with the practical argument of youths leaving his admirable Middle Class School at fourteen, with a knowledge of Science and Art "that is not at present commonly attained, in the adult stages, in public schools, or in the most expensive private schools," * there is no denying that a revolution is at hand in Education, resembling in its results that which took place between thirty and forty years ago in Locomotion. Independently of a considerable saving of time through the omission of whatever is not practically found to make children wiser or better, and through a judicious alternation of occupations, TEACHING POWER itself is undergoing a transformation which

---

* See *Society of Arts Journal*, for Dec. 17th, 1869, page 79.

reminds us of the scientific substitution of steam for muscular power ; whilst a judicious way of imparting knowledge that makes its acceptance spontaneous, is replacing the old plan of thumping it into boys' heads, somewhat as the system of smooth inviting rails replaced that of rough and resisting roads. Under such circumstances, I hope not to be far wrong if my calculations of educational progress suppose something like a railroad pace. *

THIRDLY.—A FOUNDATION FOR TECHNICAL TRAINING. It has been shown † that the scientific elements which are required for enabling a man to live judiciously, will go far towards enabling him to do his work intelligently. Any further knowledge more specially directed towards his intended occupation with which there may be an opportunity for endowing him, will of course be valuable, but here too the selection must be carefully made, for the nature of the knowledge will be of more importance than its amount. On the subject of the general value of scientific knowledge to the Workman, and of the particular value of that which being thoroughly adapted to his needs will be sure to prosper, grow, and multiply in his mind, I may be allowed to quote from the discourse pronounced at the Plymouth School of Science and Art by Bishop Temple, than whom no better authority in matters of Education could be referred to. After alluding to the somewhat exaggerated dissatisfaction with our industrial status which supervened after the Paris Exhibition of 1867, he acknowledges in the following terms the good results which may accrue from a "grumbling spirit," when it becomes the source of active exertions :—

* At a meeting at Hackney lately presided over by CHARLES REED, Esq., M.P., a new system for learning to read was propounded by Mr. SONNENSCHEIN, by which the Chairman thought that the time usually required by the children of the poorer and agricultural classes, might be shortened by 50 or even 75 per cent.

† Section V. of Memorandum.

"I have no doubt that in this way we shall really succeed in doing just what is wanted—that is, spreading over the whole of England that sort of knowledge of the principles of his work which is necessary to make a thoroughly intelligent workman, to provide everywhere the means by which any man, who has the ability to make himself master of those scientific principles on which all work must really be done, shall be able to cultivate the ability until he can rise very much above the mere routine rule-of-thumb workmen. * * * There can be no greater improvement to any one's mind than that he should thoroughly master the principles of his own work, that by which he is to live, that which is to occupy his time and his thought, that to which he is to give all the desires of his heart, the employment to which, if he is a thoroughly good workman, he would really wish to give a good and hearty service. All that really cultivates the man more almost than anything else you can teach him, for this reason; supposing you take a boy and give him a great deal of careful instruction in something which he will never require, all he has learnt of it will gradually fade out of his mind. You will find for instance, that those who have only been taught to read and have afterwards had no inducement to read, by the time four or five years have elapsed since they were at school they forget how to read. The same is the case with almost everything else you can teach. In all instances it will be found the value of it is enormously increased—I do not say it entirely depends upon it—if the future life is a perpetual commentary on the early instruction. If a man learns that which in his work afterwards is perpetually occurring to his mind, his learning won't stop when he leaves school, it will go on and on, disciplining his intellect, opening his understanding, and the chances are that he will almost invariably add to the knowledge he has at first a great deal of additional knowledge, picked up he himself cannot tell how, simply because it is perpetually present to his mind, and his work perpetually brings it back.* * * He becomes a really better educated man, his intellect is more disciplined, he is in all his ways much more intelligent. I look upon it as one of the very greatest benefits that can be conferred upon a working man that he should be enabled to cultivate his own mind; and the directest and easiest way to cultivate his mind is to enable him to acquire the principles of his own occupation."

A welcome confirmation of the value of mental culture to artisans, is afforded by Dr. LYON PLAYFAIR's lately published powerful Lectures on Primary and Technical Education. They are concluded by a condensed summary from which I feel pleasure in quoting the following conclusions:—

"That the limitation of the Revised Code to the three R's vulgarizes Education, and renders it comparatively useless for the purposes of the Working Classes."

"That common sense as well as the experience of other nations, indicates that an elementary knowledge of the principles of Science and Art involved in the occupations of the people should be introduced to primary Schools, in order to make them a fitting preparation for secondary schools."

"That a higher education, in relation to the industries of the country, is an essential condition for the continued prosperity of the people; for intelligence and skill, as factors in productive industry, are constantly becoming of greater value than the possession of native raw material or local advantages."

FOURTHLY.—MANUAL DEXTERITY AND PRACTICAL CLEVERNESS. The example of what is done at some Industrial Schools and especially those on the half-time system, shows that boys may attain a considerable amount of efficiency at manual occupations without detriment to their intellectual studies. It may frequently occur that a lad's parents cannot decide before the conclusion of his school years, what he is to be, and that he himself can scarcely make up his mind what he would like to be; but there are certain tools that every man should be able to handle, and certain trades that every artisan should know something of; and moreover there is a general cleverness of hand as well as of mind, which will help a man on in almost every occupation that choice or circumstances may lead him to adopt. It is perhaps in respect of this general cleverness, which with many is a gift of nature, but which nearly all can acquire under a proper training, of this supple readiness for adapting muscular strength and agility to purposes not included in the ordinary routine of daily work, and of this presence of mind ever available for the impromptu application of knowledge to the overcoming of unforeseen obstacles, that the workmen of some foreign countries deserve our commendation, more than in respect of any technical capability that could be named. The success of our manufacturing industry proves that English workmen possess qualities of mind, and a temperament of body, which peculiarly suit them for discharging regularly and efficiently, under the guidance of intelligent foremen and enterprising manufacturers, a definite task assigned to them in the routine of factory work, and generally involving neither science, art, nor origination. The case is somewhat different with the tradesman or mechanic, who is expected to turn a ready hand to the various duties of a complex handicraft, who must rely on his own knowledge and ingenuity for helping himself under difficulties, and who cannot always, especially in the country, wait till a brother tradesman can come and do some small fraction of his job, which protrudes beyond the regular frontier of his calling. Here we too frequently find shortcomings both as to technical knowledge, and as to the practical intelligence required for the ready application of that knowledge, shortcomings which occasion waste of expenditure, and annoyance and discomfort to the public, whilst they entail unnecessary trouble and oftentimes injury to health or limb on the artisan himself. This is by no means to be wondered at, considering on the one hand the lax and deficient system of Technical Training which has hitherto generally prevailed; deficient in the amount of knowledge imparted, and lax in its selection; and on the other hand, the absence, among a large portion of the work-payers, of that scientific knowledge which would make them alive to, and impatient of the ignorance of those whom they employ.

FIFTHLY.—ART TRAINING. The Working Man's Education is so much a battle against time, that I cannot agree with those who would make Freehand Drawing an essential item in nearly all popular education. There is a certain training of the eye to understand what is meant by a painting or a print, to appreciate the difference between a good picture and a daub, and to be capable of selecting cottage ornaments not deserving of a place in the "chamber of horrors," which I should wish every sane and seeing individual in the United Kingdom to possess; but manual ability at any branch of the Arts of Design, unless it be a natural gift in a very exceptional degree, is not to be acquired without much time and teaching. I should not be true to the utilitarian spirit in which I should wish the education of the People to be organised, if I were to recommend that every Plough-boy should be taught to draw; but on the other hand no one appreciates more than I do the value of the Arts of Design for all Artisans whose handicrafts involve origination, or a taste for execution of the designs of others. The Carpenter or the Cabinet-maker taking orders for a summer-house or a secretaire, should be able to sketch the proposed article, either in perspective or isometrically, *et sic de cæteris*; each Artificer being trained in that particular direction which he is likely to require, and receiving a quantum of artistic taste into the bargain.—Passing thence to the province of Art-workmanship, we rise by degrees till toil involves no bodily fatigue, and the Artisan is merged in the Artist. I cannot pass by this industrial region, in which we used formerly to shun comparison with other nations, without paying my humble tribute of admiration to the highly successful exertions of our friend and colleague Mr. HENRY COLE, whose system of Art Training has extended its benefits to all parts of the country, making them all contribute in return their respective quota to the national stock of artistic talent.

Perhaps some member of our Society may be inclined to ask how it is that, notwithstanding the remarkable impulse thus given of late years to the applications of the Arts of Design to industrial purposes, we read in the *Society of Arts Journal* of the 5th of March, 1869, so discouraging a report addressed to the Council by the three eminent men who had undertaken the office of judges in the Competition of Art Workmen for the Society's Prizes. As long as the supplying of Model Designs in the several branches of Art Industry had formed part of the system of our yearly special Examinations, we had had every reason to be satisfied with the ability displayed by the respective candidates in carrying them out; but the attempt to raise our Art Workmen a step higher by inducing them to be the executors of their own designs, led to the results depicted as follows in the Report :—

" In spite of the individual specimens of excellence to which we shall presently allude, we are bound to confess that the response made by Art Workmen to the Society's liberal invitation to compete for prizes offered during the last session, cannot in our opinion be regarded as satisfactory. It will be remembered that the lists of subjects proposed differed materially from those of previous years—it having been considered well, as an experiment, to test the workmen's powers in the combination of original design with skilful workmanship, and in novel directions rather than to keep them in the groove of the reproduction of the best works of the past.* * * * Whether it is that the task recently set to the Art Workmen has been beyond their present powers, or as is more probable that they look with anxiety only to what affects their regular employment, possibly, in some cases, apprehending notoriety as a fault rather than merit in their masters' eyes, certain it is that the results of their labor, taken as a whole, are not such as we had hoped for, nor such by any means, as we think would have been made by French, or even Belgian workmen, had a similar invitation been addressed to them.—We do not necessarily attribute this to incapacity on the part of our Art Workmen as executants, but ascribe it rather to their want, in this case, of the directing and sustaining power which is supplied to them, in the course of ordinary business by the superior education and attainments of their masters, and of the artists and designers, from whose drawings, models, or suggestions, they may habitually work."

Now I perfectly concur in thinking that our comparative failure in the endeavour to elicit tokens of origination from our Art Workmen, is in great measure to be attributed to the simple fact that origination is not what their employers particularly wish them to possess; and I can perfectly follow out in my mind the hint given us by our Art Judges that "notoriety," or in other words distinguished excellence in design on the part of a workman, might not exactly please a master who required merely a perfect instrument, patient, opinionless, and not possessing any merits beyond those absolutely wanted. But it is certainly not for meeting such views as these that our educational system is to be organized, and I delight in the expectation that the measures which will in all probability be adopted, for giving the children of the people a primary schooling better calculated to awaken their intelligence, will in process of time tell very effectively on their aptitude to turn to good account the facilities for artistic instruction which Mr. COLE has placed within their reach. I am not a believer in panaceas, and do not argue that Science will make a lad an artist; there is, on the contrary, a kind of science and a way of teaching it, more calculated to deaden than to improve the poetic sentiment which should pervade his appreciation of the chaste and beautiful in art; but I firmly believe that Science so selected and so taught as to raise the youthful mind towards heaven in thankfulness for God's bounties, and at the same time to awaken it to an intelligent and lively use of them, is the most effective drilling that the mind can have for the development of its highest as well as of its most practical faculties. When Science of this kind shall have opened the understanding and elevated the aspirations of the children of the People, we shall not see our soldiers hungry and shivering where a French campaigner would contrive to make a comfortable meal, neither shall we see our art workmen unable to compete with their neighbours in thought and origination.

SIXTHLY. — MUSIC AND POETRY. Among the humanising influences of which primary education should prepare, and if possible initiate the development, an honourable place belongs to MUSIC, and to this all will agree who have watched the endeavours made to promote choral societies in this country; but I would refer them to Switzerland, the Tyrol, and other parts of the continent for a more complete idea of the manner in which part singing may be introduced with success at an early age, so that it may expand into a popular resource, and become one of the best preservatives against the abuse of stimulants. The theory and practice of Music were among the favorite pursuits of my younger days, and I shall have many suggestions to offer on this subject, should I ever reach the portion of my Popular Course intended to embrace the recreations of the working classes. As regards Poetry, I have perhaps raised a smile by naming it with Music at the head of this paragraph, as one of the matters to be considered in devising a scheme of popular elementary education. Now I do not wish to train anyone to verse making, who is not born a poet; no practice is more unpractical, nor need we recur to the testimony of Horace* to know that day dreamers are likely to fall into a well. But there is an immense difference between poetizing, and possessing that faculty of poetic feeling, which gives us the perception and the enjoyment of what is noble or touching in a moral sense, sublime or beautiful in nature, masterly in art, or admirable in literature. It is this feeling which I should wish to see infused wherever practicable into the nascent energies of youth, in close companionship with moral sentiment; not didactically, for that would be a failure, but by the thousand means of gradual infiltration which a well ordered education, even that of the humbler classes, can generally command; and I should particularly wish this æsthetic development to be attended to in the rather higher stages reached by those intending to become Art Workmen. Nothing would tend more effectually to

* *Ars Poetica*, line 453.

make up for the want of a southern sky, than the brightness of conception which poetic feeling can impart.

SEVENTHLY. — RECREATION, EXERCISE, DRILLING. Whilst on the one hand no part of the proposed Primary Science Teaching should be a hardship, and many parts a treat, a favour, and a recompense, on the other hand, recreation should never be mischievous but often instructive. To enumerate the many ways in which games and pastimes in and out of doors may be made to contribute to the development of bodily strength and agility, of a quick intelligence and a ready memory, of courage, perseverance, conscientiousness, good nature and good manners, would be to add at least a score of pages to this Memorandum. Particularly deserving of notice however, is the subject of DRILLING AND MILITARY EXERCISES which, independently of their value with a view to our possessing an efficient and spirited system of national defence, more or less analogous to that of the Swiss, are susceptible of promoting very essentially the love of order, and that sense of honour which says so emphatically—" Act on the square, boys, upright and fair."

B.—APPRENTICESHIP.

No one contests the importance of the subject of Apprenticeship, or the grievous defects of the present system, both as regards the provisions of the Law, and the customary manner of carrying them out, not to say of neglecting them altogether. Yet this subject does not seem to have received its fair share of the attention bestowed of late on educational matters, and consequently there is a particularly favorable opening for those who might be able to favour us with practical suggestions on the following and other similar points :—

a. Antiquated and anomalous condition of many laws and enactments framed in industrial and social times essentially differing from our own.

b. Grounds for dissatisfaction on the part of the Master.

c. More frequent causes for legitimate complaint on the part of the Apprentice and his friends.

I have reason to believe that the latter category of grievances are frequent items among the cases brought before the Justices of the Peace, and to these therefore we might look for much valuable information in this matter. It appears that the chief causes of dissatisfaction and dispute are :— *a* That there is not any suitable criterion by which persons, and especially poor and illiterate persons desirous of apprenticing a lad, can safely judge of the abilities and trustworthiness of the Master to whom they think of confiding him. *b.* That the Master seeks too much his own profit, employing the lad at errands or other desultory work, to the detriment of his technical progress, to say nothing of his physical comfort and moral improvement. *c.* That Masters who undertake to teach a Trade, frequently know or practise only certain branches of it. As for instance Coach-makers in Country places near the Metropolis, who have the bodies down from Town and only dress them. *d.* That Masters are apt to show a kind of jealousy of communicating to a clever Apprentice, the whole amount of their own knowledge and ability.

A study of the laws and usages prevailing on the Continent will be found suggestive of improvement, and I may be allowed to introduce here a few observations borrowed from my "Letters on the Condition of the Working Classes of Nassau," printed for distribution in 1853, as a "Report to the Council of the Society of Arts."

As far as I have been able to ascertain, serious disagreements between Masters and Apprentices are less frequent in Germany than with us. This is partly referable to the prudent precaution of not signing an indenture till the intended Apprentice has had two weeks preliminary trial. Then again we find that the incompetency of the Master, which is so frequent a cause of complaint in England, is in some measure obviated in Germany by the Examination which must be undergone before an artisan can settle anywhere as Master and take charge of Apprentices. In all cases redress is facilitated by the practice of paying the stipulated sum by instalments, so that one-third or one-half of the amount stands over till the conclusion of the term. If an Apprentice has just cause of complaint, he is released by the local authorities from further obligations towards his Master, and his friends from further payment.

The question of the number of years that should be allowed for Apprenticeships in the various Trades, is one which cannot satisfactorily be discussed till many other points have been settled; but I may mention that in the Duchy of Nassau, the term for ordinary Trades, such as those of the Shoemaker, Tailor, Joiner, Baker, &c., is, or at least used to be, three years; except when the term of four years was agreed upon, without payment, the work of the Apprentice in the last year being expected to form an equivalent. Even for the more difficult Trades, such as those of the Watchmaker, Mechanician, Lithographer, &c., the term of apprenticeship did not exceed four years.

A feature of continental arrangements to which I wish to draw particular attention, are the institutions of *patronage* which in France and elsewhere, watch over the interests of the Apprentice, lend him a helping hand, and tend to keep him in the right path. It would be well worth considering to what extent similar benefits might be elicited from our Guilds and Corporations, and perhaps also from some of our Trades' Unions.

C.—Evening Classes,
Schools of Science and Art, &c.

I have adverted in Section 4, to the possibility of giving to Boys, in a Primary Education ranging from 6 or 7 to 12 years of age, a useful foundation of elementary Science, and of somewhat extending it during a secondary term of two years. The supplement of attainments acquired during this latter important period, should have if possible a tendency towards prospective future occupations, and should include elements of Science and Art selected accordingly; but at the best this would only be the prelude to subsequent prolonged special studies, for which all possible facilities and inducements should be afforded, whether or not an apprenticeship should be gone through, or some equivalent preferred. Hence the importance of developing and perfecting to the utmost the system of Evening Classes, Schools of Science and Art, and the like; not according to an uniform pattern for all localities, but with a special view to the requirements of prevailing local Trades and Occupations, whilst everywhere tuition must be carefully adapted to the capabilities and convenience of the taught, and every available resource turned to the best account. Unfortunately, owing to the extraordinary neglect of Science as a branch of public Education, which has prevailed up to the present time, we find that among the many sincere and generous friends of popular improvement in various parts of the Country, there are comparatively few who, on the one hand have sufficiently studied the general range of the Natural Sciences, to know the divergent applicabilities of the various branches of each Science, and who on the other hand have sufficiently examined the ins and outs of Industrial Life, to understand the grouping of the various categories of Youths and Adults according to their present capabilities and prospective requirements; and consequently there are comparatively few who are competent to direct to the best advantage, the organization and management of the various available means of Scientific Instruction. We may thus partly account for the educational incongruities to which I have already adverted, as having arisen since 1867 through ill-directed efforts to make up for past neglect. For instance, one sees Science of heavy calibre weighing down Schools of Design which Science of a buoyant kind might have raised to still higher success. I could name a School of Art to which has been hooked a Course of Thirty Lectures on a single division of Chemistry, that of the non-metallic elements, and would leave to the *conoscenti* to foresee the end of such a beginning. Intellectual Food fit for University Students, cannot generally speaking be expected to suit the mental appetites of Men and Lads of the Working Classes; neither are the vicissitudes of their condition compatible with Courses upon any subject, that "drag their slow length along" through a period of three years, leaving to chance which part a student may begin with, and making it very uncertain whether he can possibly attend the whole.

May we not partly ascribe to a similar cause, the exaggerated anxiety to give Working Men the benefit of listening to tip-top Professors. Scientific eminence, and even eminence in lecturing, *ex cathedra*, to advanced students, does not necessarily imply ability to render instruction easy and interesting to beginners. It is difficult for a *savant*, who through the toil of many years has arrived at a familiar acquaintance with the most recondite depths of science, to resume in thought the unfledged mind with which he began his career, to identify himself with the inexperience of his hearers, and to see the difficulties which they see. He is too apt to deal out instruction in ponderous masses, that neither fit the wants, nor the minds, nor the available leisure of industrial students, or to indulge in lofty theories, and to talk of things that are never seen, or he must at least explain Physics with Mathematics, and Chemistry with what looks to the uninitiated like Algebra. There are brilliant exceptions to this rule, but, generally speaking, the Professors best suited for Science Teaching among the Million, are men of incipient fame and moderate pretensions, less remarkable for depth of learning, than for felicity of manner in communicating knowledge, for judgment in selecting what may be most conducive to the future advantage of their pupils, and for tact and good-nature in adapting it to their present capacities.

When the management of Evening Classes is in the hands of the Professors, and what I am going to say equally applies to higher grades of scientific teaching, it is a natural consequence that the mode of dealing with the respective Sciences should be more or less regulated by the dominant abilities or peculiar predilections of the said Professors; and there is too often a tendency observable in each of them to isolate his own Science or branch of Science, upholding it as the one thing which, if thoroughly mastered, would carry all before it, and either ignoring other departments of knowledge, or at all events neglecting to make manifest the admirable results derivable from a proper interweaving and anastomizing of the twigs and branches of the Tree of Science. *

It is evident that the best remedy for many of the foregoing inconveniences, would be found in the principle of CENTRALIZATION, for help and not for hindrance, adverted to in a previous Section of this Memorandum.

---

* "However lamentable the fact, it is certain that men engaged in one branch of Science are very apt to underrate the importance of all others."— See LIEUT. COL. STRANGE'S Paper, before referred to, on the "Relation of the State to Science."

A national scheme of Education, founded on data supplied from all parts by all parties, and elaborated by the collective wisdom of our ablest Economists, would show what institutions of various kinds and degrees are required for securing our industrial progress in all its departments, and how they should be organized and conducted from an overlooking centre, in order to requite with success the exertions or pecuniary aid of their originators and supporters. Thus existing institutions would be gradually induced to purge themselves of incongruities, profiting as well as the new ones, by the prescriptions of authoritative experience, and all would fall by degrees into the regular ranks of a definite educational hierarchy.

I assume that our Educational rulers, having duly determined what knowledge, variously made up of Science and Art, is wanted by various categories of the industrial community, will wish to show by actual examples, and to encourage by liberal assistance, the modes of instruction by which such knowledge may be imparted with the best practical result at the least expense; and it is in this point of view that I hope they may find it worth their while to try on a comprehensive scale, the plan of instruction which, within the limits of my humble exertions, has met with such encouraging success. Courses of Popular Lectures and Class Lessons, embodying the kind and amount of Science required for the general purposes of Daily Life, or for the special purposes of Technical Industry, would be prepared by the most competent authors, with full sets of illustrations, and their delivery, either by the binary or the single-handed method, according to circumstances, would secure at a moderate outlay, the systematic diffusion throughout the country, of reliable and uniform instruction under perfect supervision and control.

D.—Educational Materials.

It would be a great boon to education generally, if the Central Educational Authority were to issue, or to induce and supervise the issue of various series of Diagrams, suited for popular use in Lecture Halls, Schools, Museums, &c., care being taken to unite scientific accuracy with artistic execution, and both with a thorough consideration of the class of students to be taught, the distance at which the Diagrams would be intended to be seen, whether by day light or artificial light, and the valuable resources which modern polygraphic inventions, aided in some instances by photography, present for supplying a good article at a low price. I have alluded to the difficulty which I have experienced in obtaining suitable Diagrams for my Lectures, though I use many foreign as well as English ones. Certain sets have been published deserving of much praise, but they are far from embracing in appropriate style, and on a proper scale, all the departments of popular knowledge; and as for the cheaper productions, (though still scarcely so cheap as I should wish to see them,) some of their errors and short-comings call loudly for the interference of supervisors well up in science and art, and having higher purposes in view than that of speculating on public ignorance and the prevailing want of artistic culture.

A similar opening for official supervision and assistance, is presented by the condition of our Technical Literature. There are a few thoroughly good and well-illustrated works, but their price is mostly above what working men can afford to give. It could in fact, only be brought within their reach by a certain amount of pecuniary sacrifice, which an ordinary Publisher can by no means be expected to incur, but which would most decidedly be money well spent on the part of the public purse. In the matter of minor Trade Manuals, Messrs. HOULSTON & WRIGHT, and a few other Publishers, have done very creditably as much as could be expected from unassisted speculation, as may be seen by over 30 little volumes, mostly ranging from One Shilling to Half-a-Crown, which I have got together in the Library attached to my Museum; but they scarcely can be compared with the collection of the *Manuels Roret*, published at Paris, as the result of the united labours of an association of *savants* and technicists. A few years since I procured the whole of the series then published, consisting of forty-five 18mo. volumes of uniform appearance, but different thicknesses, and of which the London prices vary mostly from Two Shillings and Sixpence, to Three Shillings and Sixpence per single volume, or Seven Shillings for the two volumes required by some of the Trades. I beg leave to draw attention to this collection, which was exhibited at Mr. DAVIDSON's interesting Lecture at the Society of Arts, on the 13th December, 1867, and is still deposited for inspection in one of the Society's Book Cases. An English series analogous to this in most respects, but printed in bolder type, very fully illustrated, and carefully adapted to the exact requirements of the rising workmen of the present day, would be a boon which their voluntary demonstrations in connection with the Club and Institute Union, have proved that they would highly appreciate, and which would amply repay in public estimation, any moderate pecuniary sacrifice that might be made for the purpose by the Government Authorities, or by the Society of Arts. I may remark, that in this as well as in almost every series of publications intended to form a homogenous group or Cyclopædia, it would be desirable that the work of many should be sufficiently regulated by single minded control, to prevent overlappings, and too great a disparity in the style and degree of condensation of the Texts, the proportionate number of illustrations, &c. This measure would not interfere with the bringing out of Class Lessons on the respective trades as explained in Section 5.

E.—Public Educational Collections.

Among the educational wants of our artisans, there are several for meeting which the most suitable provision consists, or rather might consist, of Science Museums and Art Galleries, with facilities for studying them. I will for the present, confine myself to a few remarks: firstly, concerning the inappropriateness of most of the old-fashioned collections for the purpose of Public Instruction; secondly, concerning the principles which should govern the organizing of the collections specially intended for that purpose; and thirdly, concerning the means by which the educational value of existing establishments might be improved.

The majority of the old Public Museums have been the creations of opportunity rather than of deliberate design, and consist of gifts rather than of purchases. Consequently their growth is often distorted through wealth accumulated in one department, whilst another remains poverty-stricken, or is conspicuously absent, and there is an uncomfortable incongruity about them, because they have been enriched with collections formed with different views from those with which their first nucleus was produced. Then again a large proportion of the Museum Curiosities brought home by travellers from distant climes, are sensational rather than educational; or if they supply missing links in the serial illustrations of Natural History, or serve to elucidate some knotty point of Ethnology, or Archæology, or Palæontology, they speak only to the learned, and are mute to the Working Man. Hence if we submit to an impartial analysis, the benefit supposed to be conferred on the masses by the inspection of Public Museums of the old type, we find that it consists more in rousing the intellect than in feeding it, and less in mixing recreation with study, than in preventing its being mixed with vice.

If we pass to Picture Galleries, we perceive that the principles of selection to which they owe their origin and growth, too often differ from those which would best conduce to popular improvement. Connoisseurs well up in the history of Pictorial Art, may wish to illustrate the various stages of the various Schools, and even the different styles adopted by some of the best Painters at different periods of their career. Thus Pictures may be prized and chosen for motives quite extrinsic to their artistic excellence. Sometimes even rarity makes up for want of merit. But the case will be different if we simply wish to make a selection calculated to refine the visual taste and æsthetic sentiment of the Working Classes. There is for instance, in the productions of some of the earlier Italian masters, as for example, FRA ANGELICO and the BELLINI, a style of outline which considered with due allowance for initiative genius, may be called "chasteness of contour," but which would make a false impression on the minds of most of our designers, and if imitated by them would degenerate into stiffness. Similarly the quaintness of certain specimens of other Schools would, through being misunderstood, lead to the grotesque. Again there are certain associations and conventionalities in the classical refinement of the upper classes, which rightly or wrongly, have a potent influence on their appreciation of the moral element in works of Art; whereas those whose eyes and minds are yet in a state of nature, may perceive nothing humanizing in the most artistic renderings of bodily torture, nothing edifying in the voluptuous nudities even of a TITIAN or an ALBANO, and nothing elevating in a drunken scene, even though portrayed by a TENIERS or a BRAUWER.*

Considerations of this kind are rather disheartening as compared with the sanguine expectations of improvement, which some economists attach to prospective facilities for the frequent inspection of large Science Museums and Art Galleries by the Million, and yet, behind them lies a fund of consolation and promise. Thus it will be found that Collections occupying far less space, and involving infinitely less outlay than those required for the higher purposes of Science and Art, may be very successfully instituted for the special instruction of the Working Classes, provided one adheres strictly to plain utilitarian principles, never losing sight of the proposed aim. In illustration of my meaning, I may venture a few suggestions, beginning with the Collections of a scientific character.

In organizing Museums for the Economic or Technical Instruction of the Working Classes, it is essential to start with a well defined, though somewhat elastic scheme, produced by mentally digesting together requirements, resources, space and classification. It should be somewhat elastic, because resources which one has had reason to reckon on may not always justify one's expectations, and because on the other hand certain sets of articles may come to hand, rather out of due proportion to others, and yet too valuable to be refused. But it should so far be defined, that every important requirement may be met by a satisfaction or an apology, and that a methodical and progressive array may meet

* The subjects painted by BRAUWER are taken from low life, of the most unpleasing class. From the extraordinary skill displayed in the execution, the excellent colouring, the correct drawing, and the life and character of the design, they fetch a high price; but this appreciation of connoisseurs should not secure them a place in the Galleries of the People. Again, it would be of questionable expediency to exhibit without comment, the anachronisms of costume which disfigure certain productions of noted artists, as for example, of BASSANO and even of REMBRANDT, or to display without explanation landscapes either deviating from nature, or copying from nature effects which are not what one would call *natural*, or violent architectural perspectives which are only right when the eye viewing them is at the proper *focus*.

G

the eye of the student, as he treads the path prescribed for him from the beginning to the end of his COURSE OF VISUAL INSTRUCTION. I have endeavoured to exemplify this principle in my Economic Museum,* by making the several departments of a comprehensive range of bionomic illustrations, follow each other like the chapters of a scientific work, and if I should live to see my Collection completed and transferred to London, I should wish to prepare a corresponding Text Book, rendered as readable as the subject might allow, and resembling a series of familiar Lectures of unequal lengths. It is evident that a student desirous of mastering a given range of scientific or technical information, who in successive visits to a Collection thus specially organized, attentively collates what he reads with what he sees, may enjoy as near an approximation to, and as good a substitute for, a series of oral Lessons, as can well be devised. He has not the advantage of witnessing experiments, or of being able to ask questions, but he has that of adapting his visits to his leisure, and his progression to his mental strength.**

This principle of considering educational Collections, when properly organized, as sets of Illustrations corresponding to so many special Texts, will be found to afford a most safe and useful guidance in organizing them from the beginning. A full working Programme should be prepared, similar to the detailed syllabus of a

* See Appendix No. 1.

** When first I began organizing my Economic Museum, I intended to have everywhere in convenient proximity to the various articles displayed, Instructional Labels, giving the required information concerning them; and this plan has been carried out in a portion of the Food Department, but I found that if adopted throughout, it would add too much to the difficulty unavoidably great, of packing a multitude of illustrations of all kinds within convenience of inspection, and moreover that the few persons who would take the trouble to read so much printed or written matter, would prefer taking it home with them in the form of a Hand-book. I consider that in the enlarged Food Collection instituted at the South Kensington Museum, a wise course has been adopted in displaying only condensed abstracts of pithy information, which could be printed in large type and read at some distance. Their Collection of these Labels, especially those giving food analyses, is highly instructive, and has proved of great service to many local Museums to which they have been courteously distributed, including my own.

Course of professorial Instruction, and with this in hand one will be saved many vacillations, and much unprofitable trouble and expenditure, and be greatly fortified with clearness and steadiness of purpose, both in withstanding the temptation of perplexing offers, and in obtaining the right things in the right proportions, through donation or *purchase*. I have emphasized the latter word because I have seen important Collections injured by too great a reliance on the alacrity of manufacturers and tradesmen, to come forward with their goods for the sake of the Advertisement. A moderate amount of judicious expenditure will go a great way in supplying what the Tide fails to bring; filling gaps, completing series, bringing into deserved prominence any industrial gems that the shyness or the pride of worth may have kept away, and holding up other articles to instructive censure, to which they could not of course be expected to submit themselves gratis.

The following are among the most useful categories of Popular Museums. Other denominations may be produced by cross combinations of these:—

1. Zoological, Botanical, Mineralogical, Geological, and other similar Collections, selected and classified with a view to mastering the essential elements of the respective branches of SCIENCE. Representative types should be carefully chosen, and eccentricities excluded, except where they may substantiate important points of theory, as for instance, the transitions of Species. Prints or any other available mode of illustration and analytical diagnosis, should supplement the specimens, or replace those not obtainable, or if neither can be had a Label should mark the gap.

2. A Collection more or less select representing in scientific order one or more of the great divisions of Natural History, may be made to acquire a special utilitarian character by information given on the Labels, or still more markedly by adding illustrations of processes and uses. I need scarcely name the admirable

example afforded by the MUSEUM OF ECONOMIC BOTANY at Kew.

3. A collection of not very dissimilar materials, may be classified according to localities. It may couple Commercial with Economic information, still maintaining a tincture of Science, and thus we shall have a highly instructive TRADE MUSEUM, one of the most useful varieties of which will be a COLONIAL MUSEUM.

4. A TECHNICAL MUSEUM will be provided, if we adopt the various branches of Manufacturing Industry and Handicraft for our landmarks, and group round these the raw materials with mementos of their origin, the processes or stages of manipulation with their appurtenant appliances, and the finished results with the chief variations usual in other countries. In some instances considerable interest may attach to illustrations of the technology of former times.

5. If we carefully select from each of the foregoing Museums those articles, illustrations, and explanations which have the most direct and practical influence on the Health and Comfort of the Working Man, and if we display them in common sense arrangement, that we may lead him pleasantly by the hand as we cast successively the light of Science on his wants and his resources, his causes of trouble and his means of happiness,—then we shall have an ECONOMIC MUSEUM.

I trust that these TYPES, of which it would be easy to increase the number, may serve to elucidate a principle which I well remember hearing defined by my excellent friend the late SIR WILLIAM HOOKER; namely, that differences of classification and explanation may make different Museums with nearly the same materials. This adds vastly to the amount of good which can be effected by a Central Educational Authority, retailing out intelligently what it collects wholesale on advantageous terms, or receives as legacies from temporary Exhibitions,*

* I may here mention that there is generally no better way of getting together materials for a Special Museum, than to organize a corresponding temporary Exhibition.

or picks from the unused stores of long established Institutions. Another circumstance that brings into relief the advantages to be expected from a well-organized system of distributions and exchanges, is that Collections for the instruction of the People, whilst they need not be very large, ought to be very numerous, and should, generally speaking, be constituted everywhere on nearly the same principles. When the central authority shall have arrived at a clear view of the general educational wants of our Artisans, and of their particular requirements in the form of Public Museums, more or less corresponding to the above types, or embracing two or more of them, there will be no difficulty in preparing a number of DRAFT PROGRAMMES, for being issued to localities that may show themselves deserving of help, by their readiness to help themselves. By means of these schemes, discussion will be kept to the purpose, negotiations rendered easy, and concerted action made to yield its best fruits.*

The organizing of ART GALLERIES, which would generally be associated with Collections of the foregoing kind in all well conditioned Popular Museums, involves less variety of purposes, and the simple rules which should govern the selection, are mostly derivable by a little reflection from what has already been said. As regards the general mass of visitors, the chief object would be to raise the standard of visual taste and æsthetic sentiment, but not unaccompanied with those moral aspirations which form the noblest attributes of high Art. Nor should any indulgence be shown to that promiscuous admiration of all that bears a noted name, which in Art as well as in Literature too often ap-

---

* The subject of the best appliances and contrivances for displaying and preserving the contents of a Museum and for rendering them easily available for lecturing purposes is, as a matter of advice and assistance, little inferior in importance to obtaining and classifying them, and a central permanent Exhibition devoted to this speciality might render great service.

proaches to "Hero Worship."* These considerations are not changed but concentrated as regards the Artisans whom we particularly desire to instruct, from the Joiner or Plasterer who should learn to appreciate elegance in a scroll or a rosette, up to the Art Workman or the Working Artist. Every branch not only of Drawing and Painting, Sculpture and Architecture, but also as far as practicable, of Art Workmanship, and of the Arts which produce a multiple of copies, or Polygraphic Arts, should be illustrated by examples which in Galleries on a full scale would ramify into shades of artistic character, but in small ones would be confined to the most representative types. One art should everywhere prevail, that of affording much instruction for a moderate outlay. Nearly the same notions may be conveyed to the minds of the uninitiated by copies and casts as by originals. Fac-similes of Works of the highest merit may thus be secured, and a whole department of Art may be taught for what a wealthy amateur spends on a single Piece.* *

The popular utilization of existing Science Museums and Art Galleries, besides being furthered as a matter of course by any measures for remedying defects and carrying out principles of the nature of those which I have pointed out, may be rendered more convenient and effective by devices such as the following. One might anticipate excellent results from visits of groups of Working Men and Youths to Establishments like the British Museum and National Gallery, under the friendly guidance of persons competent to analyse their intellectual wants, to scan their capabilities, to

* I will confess that I was not so delighted as some of my friends, when I heard that over a thousand originals by the hand of TURNER, were stored away in certain depositories of the National Gallery Building, and might be available for distribution to local Institutions.

* * Appendices Nos. 3 and 4 to this Memorandum, show by way of suggestion two rough sketches of Institutions of different calibres, designed to place Science and Art Instruction within the reach of the Working Classes, with appropriate accessories.

select in proper order the objects which may serve to impress useful information on their minds, and to impart this orally in an agreeable manner. Unfortunately, the examples of the union of all these desiderata have hitherto been few and far between, and I am not sanguine as to future prospects in this direction; for the man of abilities who is zealous enough to make himself the Artisan's *cicerone*, is seldom discreet enough to see that his own favorite topic, which may be the deciphering of a cuneiform inscription, or the history of a Mummy, is not the thing for the Workshop.*

Much good may be done in Natural History Collections, by appending Labels conspicuous for their size or particular colour, to those objects which possess special interest in an Industrial point of view. When thus brought into sufficient prominence, those objects will be even more interesting for being surrounded with others illustrating the Order to which they belong, than they would be if picked out and grouped by themselves. The information concerning them might be partly condensed on the Labels, but I would chiefly rely on special Handbooks or "Visitors' Companions," published in the same colloquial style as the Text Books for special Museums referred to above.** Truly valuable for the utilization of such an Establishment as the British Museum, would be a series of cheap

* In certain Institutions, it answers very well to include in the staff, Demonstrators qualified to take visitors round in groups, entertaining them with oral instruction; but various difficulties prevent this from becoming a general practice.

* * The idea of a *Vade Mecum* familiarly explaining the features of interest of a public collection, is by no means a novel one. Since I first advocated that particular kind of Guide to Knowledge, chance has placed in my hands a curious old specimen, so suggestive that I wonder the plan has not long since become more general. It is an 18mo. of between 160 and 170 pages, published as far back as 1805. The title page is as follows:—"Visits to the Leverian Museum, containing an account of several of its principal curiosities both of Nature and Art, intended for the intruction of young persons in the first principles of Natural History." At the end of this little Volume is an explanatory List of Books designed to make knowledge agreeable, blending it with that utilitarian morality, the appreciation of which we have now so much difficulty in rousing again from its slumber.

Guides of such a character, each dedicated to a particular trade, occupation or pursuit, or to a group of them. Thus the Furrier, the Druggist, the Lapidary, and a score of others, including pursuits of a high level, would each be led by the hand to visit intelligently with no loss of time, every department in which objects, whether many or few, could be made to yield useful and entertaining knowledge specially addressed to their respective callings.

F.—Technical Examinations and Certificates.

Among the technical institutions or customs of the Continent, there is scarcely any one of which the introduction into this country would require a more careful adaptation, but which if once satisfactorily introduced, might be productive of more useful results, than the system of Examinations in technical knowledge combined with manual ability, through which it is or has been the practice in certain countries, to sift the qualifications of matured Apprentices aspiring to become accepted Journeymen, and of experienced and clever journeymen wishing to take rank as Masters in Trade, to set up shop on their own account, and to become the authorized instructors of another generation of Apprentices. I may be allowed to borrow from my "Letters on the Condition of the Working Classes of Nassau," the following account of the manner in which the Examinations in question used to be conducted. It has already been quoted in the numbers of our Journal for January 3rd and 10th, 1868.

" At the expiration of his term, the Apprentice must furnish proof of the extent of his acquirements, by executing some appropriate piece of handiwork, in the presence of the official judges of the Trade, forming a kind of jury, which from its usefulness deserves some attention. Every three years the masters in each Trade residing in a district, or in a group of districts if the trade is a scarce one, assemble to elect or re-elect three representatives for the purpose of examining the Certificates, and of testing and recording the abilities of industrial Candidates. Such is the Board of Examiners, which we now find sitting in judgment on the merits of the young artisan anxious to emerge from his apprenticeship, and which we shall meet with again in a further stage of his career. If the Examiners are not satisfied with the young man's performance, he must find means of improving himself within half-a-year, against another trial; if on the contrary, they are well-pleased, he obtains his certificate of *Gesell*, or Journeyman, and sets out for his travels, or *wanderschaft.*"

As I do not see how the practice of *wanderschaft* could be satisfactorily adopted to any extent as a regular part of an artisan's training in this country, I omit the account given of it in my "Letters on Nassau," and proceed at once to the second or major Examination, which the Journeyman must pass to become a Master. * * * "He is required to accomplish single-handed, for strict inspection by the Board of Examiners, some model piece of workmanship, sufficient to show, not merely a moderate amount of skill, as when he was a candidate for a journeymanship, but his thorough knowledge of the *arcana majora* of his calling. If he can follow up the display orally with theoretical evidence, he is entitled to be admitted forthwith to the honourable company of the Masters of the Trade."

Such is the system which, with local modifications, has largely prevailed in Germany, and has worked well, as may be seen by the following extract of a Letter which I received a few years ago from a very intelligent manufacturer of Oschersleben, in Prussia :—

"I enclose an abstract of the law in Prussia concerning the exercise of Trades and Handicrafts. * * * It works admirably, and the public are certain to be served by men who understand their Trade. In your country any person can commence a Trade, whilst in ours he must first show that he has the necessary capacity and knowledge for its successful exercise."

Yet in Nassau and elsewhere this system has been abandoned, and I have been told that in many parts of

Germany where it is still kept up, there is a growing inclination to drop it. The reasons however are tolerably obvious, and do not at all affect the good which might be derived from introducing Trades' Examinations in England in an altered form. The chief reason is that in the countries in question, these Examinations have been compulsory, and coupled with obnoxious restrictions to the free exercise of handicrafts. Nor could one always rely on the impartiality of the Trades' Juries, especially as regards the admission of Journeymen to Mastership. Suppose for instance, a clever Journeyman Shoemaker, who has lately been working at one of the most fashionable establishments of Paris, and who now wishes to set up on his own account in his native German Town, with an ill-disguised intention of importing the goods of his late employer, with whom he has entered into an understanding that anyone can guess. Now it would be scarcely consonant with our knowledge of human nature, to imagine a conclave of three master shoemakers acknowledging the merits of this Frenchified intruder, and the chances are they will find some good reason why he should stick to his last for a few years longer. Again it is obvious, that any arrangement like that above described, tends to isolate the members of the several trades in the several Towns or Districts, grouping them into close Corporations, all probably rather antipathetic to innovation and progress, but nevertheless some more enlightened and more clever than others. Consequently they would measure technical abilities, whether mental or moral, by different standards of merit, and a Certificate of competency for Journeymanship or Mastership obtained at one place, would have a different significancy and real value from one obtained at another place. The remedy for this would be CENTRALIZATION. The remedy for the other defects would be to make the system *voluntary* instead of *compulsory*.

Might we not, without binding ourselves to the German rules of Journeymanship and Mastership, adopt the idea of having definite grades in all the ordinary handicrafts, such as those of JOURNEYMAN or COMPANION, and MASTER or TRADESMAN. The Apprentice in any given Trade, (say for example that of the Dyer,) would after the conclusion of his Apprenticeship, be invited to pass an examination, not only in the various departments of elementary and applied knowledge which I have described elsewhere as collectively constituting the Dyer's Art, but also in the practical performance of a few typical dyeing operations. I will not stop to discuss the means by which such examinations might best be rendered practicable and convenient, or the manner in which the authority of a Central Educational Power could best be exercised to render them reliable, and to ensure a uniform standard for each Trade throughout the Country. As for the inclination of the competent Apprentice to present himself to be tested, this would mainly depend on the inclination of Employers to avail themselves of the criterion which this test would afford, and to engage a young man whose competency as Journeyman or Companion was proved by a satisfactory Certificate, in preference to one without it. Seeing how much good will our Master Tradesmen display in attending to their own interests, and trusting to the progressive diffusion of Science for enlightening them in the same degree that it will enlighten the paying public, I see no reason to doubt that the possession of such a Certificate would add a good percentage to the marketable value of its owner, and nothing more is required to make the Examinations in question grow popular and flourish. These are the MINOR EXAMINATIONS.—MAJOR EXAMINATIONS for entitling a Journeyman to take the title of Master, would be organized on the same voluntary plan, with the same guarantee of reliableness, and consequently with the same intrinsic value, and the same claims to acquire practical value through the appreciation of all concerned. In pursuance of the principle of allowing to Trade the utmost freedom compatible with public security against ignorance and deceit, the JOURNEYMAN might set up shop, and employ those who might be willing to serve him, but a blank escutcheon over his shop door or window, would declare his small pretensions; whereas the MASTER TRADESMAN would glory in an appropriate badge, which would betoken his being in possession of a MAJOR CERTIFICATE or DIPLOMA. It is not always easy to prescribe limits to the liberty of the ignorant to deceive the more ignorant, without being accused of interfering with Freedom of Trade; but I certainly think it might be made illegal for any person not possessing a proper Certificate, to receive an indentured Apprentice.

G.—THE CENTRAL INDUSTRIAL COLLEGE, OR PANTECHNIUM.

What Evening Science Classes of various descriptions do for the Apprentice, they may of course continue to do for the Journeyman, affording him with equal propriety the somewhat higher intellectual sustenance his mind has become capable of digesting, and which, supposing the above plan of Technical Examinations in two degrees were adopted, he would want for passing the higher degree, and becoming a Master Tradesman. But with the fullest appreciation of the good which may be effected by such means, strengthened as they are already by the laudable exertions of the Science and Art Department, and as they might perhaps further be by some of the measures suggested in the foregoing pages, yet it is obvious how inadequate they necessarily must be for supplying the higher technical culture which many Trades and Handicrafts require; whether they have to borrow their chief support from Science or from Art. Hence the clamour for Industrial or Technical Instruction to which our Society has responded with so much alacrity. Nothing is more desirable than that we should encourage and foster with all the means in our power, the generous willingness which prevails in several parts of the kingdom, to work hard and subscribe largely, for establishing local Technical Colleges and Science Schools; and yet it might prove a real misfortune if these Establishments were so wealthy and independent as to dispense with Government aid; for as I have said before, the exercise of a discreet and friendly control, to which a central educational authority would be entitled through conferring substantial benefits, would be the best means for combining the efficacy of harmonious action, with the vigour of autonomy.

It would be premature to guess the course of events; but I hope it will ere long be acknowledged that in order to be able to dispense to the Industrial Community at large, and more particularly to local Technical Colleges, the greatest amount of practical assistance, a CENTRAL INDUSTRIAL COLLEGE, POLYTECHNIUM, or PANTECHNIUM, should be established in or near London, on such a footing that it might be their obvious interest to become its affiliates. I have already alluded to a plan for the creation of such a College, which I suggested as far back as 1859, to my talented friend Mr. JOHN SCOTT RUSSELL. I had the satisfaction of finding him as thoroughly convinced as I was of the desirableness of establishing an institution where the standard of technical progress might be conspicuously raised, and which might serve as a centre for diffusing those scientific and artistic attainments, in which we knew that foreign workmen were beginning to make rapid strides. I subsequently offered to subscribe £1,000 if my friend should see his way to rousing public opinion in favour of the scheme, and thus preparing its accomplishment on a national scale. Unfortunately the apathy not to say antipathy with which Science was then looked upon by a large proportion of our educationalists, was not to be overcome till the disparity of technical progress in this country and abroad, had sufficiently accumulated up to 1867, to produce an educational crisis. Thus my small pamphlet entitled "Notes on the Organization of an Industrial College for Artisans," printed for private circulation towards the end of 1851, shared the fate of better seed similarly sown before the right season by Dr. LYON PLAYFAIR and other precocious advocates of what now no one contests. From that Pamphlet the following suggestions are partly borrowed.

The PANTECHNIUM or CENTRAL INDUSTRIAL COLLEGE,

should be so constituted, and on such a scale, that our local Colleges would naturally look up to it, and that even the best continental Institutions of the kind should not look down upon it.* It should show judgment, forethought, and economy in every feature, and be a model institution in a model building, attracting from all parts the best abilities, and radiating in all directions the most genuine improvement. It would embrace Science and Art, and the applications of both to technical processes and practical workmanship, uniting with the most suitable arrangements for sedentary study, spacious technical laboratories and workshops. A system of Technical Examinations, carried out on uniform principles in all parts of the country as suggested above, would materially facilitate the sifting of Candidates for admission to the Central College, as well as the fair appropriation of numerous Exhibitions or Scholarships. Conspicuous among the rewards to be adjudicated to the most deserving students at the close of their one, two, or three years' stay, according to the nature of their studies, would be Diplomas of Competency to become teachers at Colleges or Technical Schools including the highest grades and latest improvements of the respective Trades. More or less analogous to these Diplomas, whether embracing collectively the whole of the requirements of a given Trade, or separately recording distinct attainments, would be those for Foremen or Managers of Industrial Establishments. Some of the more complex industrial occupations might possibly be best perfected by special schools, as examples of which, I may point to the School of Mines, the Agricultural College at Cirencester, the Continental Schools for Forestry, for Builders, &c.

* For an account of the "Technical Institute" at Zurich, the "Gewerbe Schule" of Berlin, and the Technical Universities of Carlsruhe and Stuttgardt, see page 40 of Dr. LYON PLAYFAIR'S Lectures on "Primary and Technical Education," (Edinburgh: Edmonston & Douglas, 1870) and J. SCOTT RUSSELL'S noble work "Systematic Technical Education for the English People," (London: Bradbury, Evans & Co., 1869.) The whole of Chapter VIII is devoted to Technical Colleges.

A prominent feature of the PANTECHNIUM, and one calculated to interest the public at large, would be its indispensable NATIONAL INDUSTRIAL MUSEUM, which should aid in the formation of analogous local Museums, not merely as serving as a pattern, but also by the various means of active assistance which have been described elsewhere, as benefits to be conferred by the judicious, and not very expensive exertions of a Central Educational Authority.

Where should the Central Industrial College be located? This question, one of the first which present themselves to the mind, is one of the most difficult to solve, because it is only too easy to find fault with almost every site likely to be suggested. The most likely of all is a part of the vacant site at South Kensington, at the disposal of Her Majesty's Commissioners for the Exhibition of 1851, and there is no doubt that for the student in general, and those in Art Workmanship in particular, the proximity of the collections which are or will be congregated in that quarter, would be of great value; but on the other hand we must remember that the inmates of the PANTECHNIUM, though the very cream of the Working Classes, would be and ought to remain for the most part *bona fide* Working Men, who even as trained and certificated Professors, would not put on gloves to wield the sledgehammer. I fear that the time spent by them amid the splendour and gaiety of one of the most fashionable parts of the Metropolis, might be little fraught with present satisfaction, and less conducive to their future contentment, and whilst I should be anxious to spare the aristocratic inhabitants of the neighbouring mansions some real and much imaginary annoyance, I should be still more anxious to see our Artisan Students located where they might feel themselves at home, and be at liberty not only to dress and live according to their means, but also to accomplish without obstacle, the practical union of Work and Study, which forms so interesting a feature of that admirable Institution, the CORNELL University.

It would not be difficult to find urban and surburban sites infinitely preferable to a choice quarter of the West End in these respects; there would be space for *ateliers* on a full scale, where smoke and hammering would not be pronounced intolerable nuisances, and where communication with South Kensington would be rendered sufficiently easy by means of modern locomotion, to secure the few extra resources which the College might not itself afford. There is one locality where, if sufficient space could be secured by parliamentary means, the PANTECHNIUM might be most eligibly located, and brought into close association with a very promising popular institution, for mutual benefit. That institution is the EAST LONDON MUSEUM, which thanks chiefly to the untiring energy of our excellent friend MR. ANTONIO BRADY, is now being erected in Bethnal Green. I need not dwell on the instructive lessons of manufacturing and commercial industry which that end of London would afford, or on the cheap resources which it would present to a concourse of intelligent and thrifty Working-class Students, or on the satisfaction with which their advent would be hailed by a needy and almost desponding population. I must confess however that I am not personally acquainted with the locality; nor am I sufficiently aware of the plans which the Science and Art Department may have in view in reference to the Museum in question, to be quite sure that the proposed combination would be a harmonious one. It is of paramount importance for the permanently successful working of the various educational institutions with which we hope to see the country enriched, that their distribution should be made a matter of long-sighted study, that those should be united which may help each other, and those kept apart of which the union might engender friction, or be a practical incongruity.

H.—The Educational Scheme of the Future.

I have already touched on the prejudice partly well founded and partly groundless or exaggerated, which prevails in this country as to Centralization, and expressed a hope that a clearer sense of duties and interests, will bring about such an understanding between the Government and the People, as may render perfectly natural and unobjectionable, that the former be entrusted with that amount of stimulating and directing influence, as well as devising and creating power, without which local energies would *here* remain dormant, *there* be wasted in spasmodic exertions, *elsewhere* be playing at cross purposes, and *everywhere* be leaving links disconnected and gaps unfilled. I do not however conceal from myself that the task of systematizing our Educational Institutions is by no means an easy one. In a country where so many already exist, and claim respect for their good points or veneration for their past services, where the ground is held by deep-rooted customs and prejudices, and where vested interests are ever ready to defend their own, the most desirable reforms might be jeopardized by urging them on too fast, and it would be worse than futile to attempt any sudden transition to a perfect state of educational organization. But what can be done is, after having collected and sifted the best information and suggestions, to chalk out a comprehensive educational system designed to meet the following desiderata: Firstly,—It should blend as much as possible those features which would everywhere be considered types of excellence, with those specially required to meet the idiosyncrasies of this country. Secondly,—Interference with existing interests should be avoided, existing institutions utilized, and voluntary aid turned to account as far as consistent with the general weal. Respecting existing interests, I may remark that certain necessary portions of our future educational system, which would raise a storm of antagonism if announced for immediate execution, might gain easy acquiescence on an understanding that their realization would be postponed; for men are more ready to be patriotic at the expense of their successors than at their own. Thirdly,—Public opinion should always be treated with the greatest respect, but its verdicts must be accepted with discrimination. It is subject to paroxyisms of energy which require clever management, and fits of reaction or torpor which require patience. On the whole it has advanced amazingly as regards educational views within these last twenty years. From this change its future progress towards thoroughly rational principles may be fairly computed, and a prospective scheme should be framed accordingly. Fourthly,—A sufficient amount of elasticity should pervade the whole system, to allow of its being adapted to unforeseen circumstances, and a large margin should be left to the discretion of a competent and trustworthy Executive.

Such are the principles on which I hope to see the eminent men in whose hands lie our educational destinies, marking out the "SCHEME OF THE FUTURE," that immediate steps may be taken for realizing what can at once be realized, and that all patriotic endeavours may tend henceforward to concordant progress in the right direction.

In the thirty-three years which have elapsed since my return to England after imbibing a love of method through a continental education, I have seen millions expended without forethought or concerted action. Railways and other public works have been constructed, buildings erected or altered, and establishments instituted which have not only not produced the amount of good that might have resulted from their being made parts of connected schemes, but have in many instances been found after a certain number of years, to act as *vetos* standing in the way of better things. Now I sincerely trust that our future National Educational System, for preparing which the present tone of the public mind offers so peculiarly favourable a juncture, will not be made up of desultory and incongruous bits, but that first of all (I must be allowed to repeat this essential principle), a comprehensive and homogeneous plan will be devised, discussed, settled, marked out, and its realization *begun* at once; that every encouragement will be given and every reasonable freedom of action will be allowed to Individuals and Bodies willing to co-operate with Government in the great undertaking, subject nevertheless to such conditions as may prevent cross-working, and make every particular part fit in with the harmonious conception of the whole. Thus by degrees a well concerted scheme may step by step become a satisfactory reality, and the present age may before its close, see in full operation an educational machinery worthy of this enlightened country. It was thus that my Father saw rising on the banks of the Potomac towards the close of the last century, a few buildings apparently disconnected, but which through the progressive development of a plan judiciously marked out in the wilderness, have become concordant features of the great City of Washington. *

---

* For Supplementary Notes, see Appendix, No. 5.

APPENDIX No. I.
(See Page 11.)

SYNOPSIS
OF THE CHIEF SERIES OF ILLUSTRATIONS DISPLAYED IN THE TWICKENHAM ECONOMIC MUSEUM.

CLASS I.—BUILDING DESIGNS.*

Preliminary Illustrations. Homes of the People in former times and foreign countries—Designs, and Models to scale, for healthy and comfortable abodes in town or country.** *Cités Ouvrières.* Model Villages, &c.—Designs for Establishments of all kinds, for public advantage or for the benefit of the Poor.—Illustrations of the best construction of the several parts of buildings.

CLASS II.—MATERIALS FOR BUILDING AND FOR FURNITURE.

Educational and Commercial Series of the chief varieties of Stone.—Mortar and Cements.—Artificial Stone.—Bricks and Tiles.—Paving Materials.—Roofing Materials.—Wood of various kinds for building, furniture, &c., with botanical illustrations. — Processes for preserving and rendering fireproof. — Varieties of Window Glass, and processes of manufacture.— Materials and appliances for house-painting, whitewashing, staining, and other decorative purposes.—Paper-hangings.

CLASS III.—FIXTURES, FURNITURE, AND HOUSEHOLD UTENSILS.

Educational Series of the Metals, from the Ore to the prepared state. Alloys. — Building Ironmongery, including fastenings of every description, and a priced series of articles required for a Model Dwelling.—Appliances for water-supply. Filters.—Appliances for warming, cooking, &c. (Grates, Stoves, and other cumbersome articles, are collected for a more convenient display, in a special part of the building.)—Select examples of household ware and furniture, showing the various ways in which things are made in this country and abroad, and the relative advantages.—Articles supplied by the Furnish-

* This class is calculated to afford, in conjunction with the following ones, every information and guidance to persons desirous of erecting improved dwellings for the Working Classes.

** Much attention has been bestowed on this Department, which now comprises a numerous selection, classified and catalogued.

ing Ironmonger and Cutler, the Gasfitter and Lampist, the Earthenware, China and Glass dealer, the Turner and Brushmaker, the Upholsterer, &c.

CLASS IV.—TEXTILE MATERIALS, FABRICS AND COSTUMES.

Preliminary Illustrations. Dyeing, &c.—Materials as produced, and in various stages of preparation for the loom.—Matting, Drugget, Carpets, and the like.—Textile fabrics of all kinds for household purposes, and for apparel.—Trimmings. Accessories of Dress.—Priced List of Outfits, suited to the various requirements of the Working Classes.—Select articles of Hosiery, ready-made Clothing, and Coverings of every description, for men, women, and children, with illustrations of the manufacture of Waterproof Articles, Hats, Shoes, &c.—Clothing for infants used in several countries of Europe.—Specimens, models, and prints of the Costumes of various parts of the world.

CLASS V.—FOOD, FUEL, AND OTHER HOUSEHOLD STORES.

Preliminary Illustrations of the philosophy of Nutrition.—Proximate constituents of Food.—Dietaries.—Unwholesome Food. — Adulterations. — Preservation of Food. — Culinary Science.—Food Auxiliaries: Water. Salt.—Food supplied by the Animal Kingdom, including mammalia, birds, fish, shell-fish, eggs, milk, &c.—Food supplied by the Vegetable Kingdom, including: cereals; leguminous seeds, or pulse; roots, and bulbs; vegetables of which the stalk, leaves, or tops are chiefly eaten; fruits; flowerless plants available for food; secreted and extracted products; cakes and confectionary; condiments. Narcotics.—Cocoa, coffee, tea, and their substitutes.—Fermented and distilled liquors.—Refreshing drinks.

Fuel of all kinds, and materials for ignition.—Lighting materials.—Stores for washing, cleaning, scouring, &c.*

CLASS VI.—SANITARY DEPARTMENT.

Public Works for protection against inundations, &c., for the improvement of unhealthy districts, for water-supply, sewerage, &c.—Appliances for the ventilation of dwellings, for preventing inconvenience from damp, smoky chimneys, noxious effluvia, &c. Disinfectants.—Appliances for Hygienic Exercises.—Baths and other means for promoting a healthy condition of the skin.—The Hygiene of Dress.—Nursery appliances.—Orthopedic apparatus. Crutches. Artificial limbs.

* It would be optional, in adopting the general plan of the present classification, to form a distinct class of ' Materials and Appliances for Warming, Lighting and Cleaning.'

Dentistry. Means of relief for deafness, defective vision and weakness of the eyes. Means of occupation for the blind, and specimens of their work.—Appliances for the sick-chamber. Means of comfort for invalids.—Household remedies. Educational series of drugs commonly used, with botanical illustrations. Medicinal herbs.—Popular patent medicines, with their composition. Drug adulterations. Injurious articles likely to be mistaken for medical substances.—Antidotes and treatment for poisons and venoms. Prints of poisonous plants, venomous reptiles and insects, noxious fish and mollusca, animals in a rabid state, &c.—Means of safety from, and destruction of beasts of prey, house and field vermin, &c.—Means of safety or relief from the effects of excessive heat or cold, from asphyxia, drowning, shipwreck, lightning, fire, accidents, and discomforts in travelling by railway or otherwise, accidents to which children are liable, &c.—Prevention of the injuries and diseases which specially attach to industrial occupations.*—Articles of which the manufacture is injurious. Eligible substitutes.

CLASS VII.—HOME EDUCATION.—SELF INSTRUCTION.— RECREATION.

Home education of children. Instructive toys. Educational prints and cabinets.—Adult self-instruction and scientific recreation. Illustrations of the various Sciences. Cheap and serviceable apparatus, for in or out-door studies. Formation of Herbaria, &c.—Illustrations of the Arts of Designs and Polygraphic Arts, showing the processes employed, and the respective results.—Instruction in the principles of taste, in outline, colour, and subject; select prints, figures, and other articles for Cottage Decoration.—Music, vocal and instrumental, in its popular applications. Modes of instruction and notation adopted in various countries.—Rational Games. Gymnastic Exercises.

CLASS VIII.—MISCELLANEOUS ARTICLES, NOT REFERABLE TO THE FOREGOING CLASSES.

Scientific appliances for household use, including clocks, barometers, thermometers, scales and weights, measures, &c.— Stationery in all its departments.—Miscellaneous Household Requisites. Toilet articles.—The Housewife's Work Box. Female handwork of all kinds.—The Cottager's and Emigrant's assortment of Tools for carpentering, shoemaking and farriery. Garden and field implements.—Seeds for horticulture and small husbandry. Resources for barren localities.—Appliances

* This important subject was brought under the notice of the Society of Arts in 1854, under the name of Industrial Pathology.

for locomotion and the conveyance of burdens. Contrivances of all kinds for lightening labour.—Special requirements of Travellers, Emigrants, &c. Self-help for emergencies.—Samples of Museum Fittings and Appliances, with estimates for the use of persons desiring to form Economic Collections on any scale of development.

CLASS IX.—THE ECONOMIC LIBRARY.

Books, pamphlets and documents, (British and Foreign,) selected and arranged for convenience of reference in matters of domestic, sanitary, educational, and social economy, and practical benevolence; especially intended for the use of persons engaged in the organization of Provident and Charitable Institutions, and of Clergymen, Medical Men, Schoolmasters, and others entrusted with the bodily welfare, or intellectual guidance of the People.

Part I of a Descriptive Catalogue has been printed for private circulation, under the title of "Handbook of Economic Literature."

The value of sound principles of Household and Health Economy for all classes of society, is rapidly gaining a due appreciation both in this country and abroad, and it is to be hoped that the effective means for imparting them exemplified by the Twickenham Museum, will obtain ready and general adoption. Of course, Economic Collections may vary infinitely in scale and character; they may be blended with almost every rational device for popular recreation, and made a place of resort for the sight-loving as well as for the studious portion of the public; they may be established on purely philanthropic or on partly commercial principles; they may respond to the special pursuits of any class of men, agricultural, mining, manufacturing, commercial, seafaring, military, etc., and be made to represent the requirements and resources of any race, climate, or locality. In certain seaports, Emigration will claim a distinct collection, and suggest a most interesting one.—Mechanics' Institutes, and other Associations for self-improvement, should allot a space to the illustrations of the Science of Common Life, for which tradesmen will readily supply samples, whilst some of the members will take charge of the manual, others of the intellectual labour.—Educational establishments, even down to the village school, should get up their appropriate cabinets of useful and entertaining objects, on nearly the same co-operative plan.

In all these cases, there will be an interesting scope for the exercise of discernment and forethought in the selection of the series to be included, and in the utilization of every available resource for illustrating them. These considerations, as well as the amount and nature of the disposable space, will in some measure, determine the arrangement of the articles, and even of the departments; but generally speaking, the classification above given, which is the result of much practical experience, will be found to afford convenient and reliable guidance. An important point to be borne in mind is that, in each department, the illustrations are to be, as nearly as possible, those which a popular lecturer would wish to place before his audience, and that the pith of what he would say is to be appended, as far as space may allow, in the form of readable and familiar labels. Specimens of these, together with drawings and estimates of cheap and compact fittings and appliances, many of which are of peculiar contrivance, will be forwarded to persons who may be prepared to organize economic collections for public advantage. In certain cases duplicates will be distributed and other assistance afforded.

APPENDIX No. II.

(See Page 20.)

The following is the substance of the Programme of the Popular Lectures described in SECTION 2.

SCIENCE MADE EASY.

FAMILIAR LECTURES

ON THE

APPLICATIONS OF SCIENCE

TO

THE REQUIREMENTS OF DAILY LIFE,

Offered gratuitously to Institutions established for the promotion of

POPULAR IMPROVEMENT.

FOURTH SEASON, 1869--1870.

This connected course of Lectures is intended to unite in an entertaining form, those elements of practical knowledge which most essentially tend to the promotion of health and comfort, and constitute the ground-work of DOMESTIC AND SANITARY ECONOMY. It has been carefully prepared by MR. T. TWINING, who supplies Illustrations of every kind from his ECONOMIC MUSEUM, and undertakes the whole of the expense involved in the delivery of the Lectures in London or the Suburbs, on condition that the public be admitted FREE; the recreation and practical benefit of the Working Classes being the sole object in view.

Reader:
MR. WILLIAM FREEMAN,
Curator of the Twickenham Economic Museum.

Demonstrator:
MR. WILLIAM HUDSON,
Of London University, Secretary & Chemical Superintendent of the Museum.

Applications for the Lectures may be addressed to T. TWINING, ESQ., *Perryn House, Twickenham.*

SYLLABUS.

LECTURE I.
INTRODUCTION.

Definition of Domestic and Sanitary Economy, or the Science of Common Life. Particular importance of its study for securing health and comfort. Necessity for a preparatory knowledge of the Elementary Sciences on the application of which it is founded.

MECHANICAL PHYSICS.

The three conditions of matter: solid, liquid, and gaseous. Distinctive properties of solids: compactness, porousness; comparative hardness; brittleness, toughness, malleability, ductility, tenacity; flexibility, elasticity; sonorousness; opacity, translucidity, transparency. Crystallization. Outward forms of bodies.

Gravitation. Weight; specific gravity of solids. Comparative density of liquids, areometers. Weight of Air. Comparative density of gases, balloons.

Inertia; momentum; centrifugal force.

LECTURE II.
MECHANICAL PHYSICS, *continued*.

Gravity as a motive power; increasing velocity of falling bodies.

Theory of the centre of gravity, illustrated by numerous examples of its application.

Important mechanical law of the inverse ratio of power and speed. The lever; the pulley; the inclined plane; the wedge; the screw; the roller, and the wheel and axle.

Rationale of wheel carriages.

LECTURE III.
MECHANICAL PHYSICS, *concluded*.

Notions of Aerostatics and Hydrostatics. Phenomena connected with the pressure of air and of water, and rationale of their most useful and interesting applications; including the diving bell, the syphon, the common and the forcing pump, the barometer, &c.

Notions of Acoustics. Production and transmission of Sound.

LECTURE IV.
CHEMICAL PHYSICS.

General considerations concerning Light, Heat, Electricity and Magnetism.

Light: its production and transmission. Reflection, refraction, and their practical applications. Decomposition of the solar ray; phenomena of colour.

Heat: Expansion of solids. Expansion of liquids; thermometers. Expansion of gases; artificial ventilation. Mode of heating liquids. Transmission of heat, by contact and by radiation; comparative conductibility and radiating power of various substances and surfaces. Changes of the condition of matter produced by heat or cold. Interesting and important phenomena connected with the boiling of water, and the generation of steam. Production of cold by evaporation.

LECTURE V.
INORGANIC CHEMISTRY.

Chemical attraction or affinity. Difference between mixture and combination. Synthesis and Analysis.

Simple or elementary bodies. The most important of the non-metallic elements, viz.: Oxygen, Hydrogen, Nitrogen, Chlorine, Carbon, Sulphur, Phosphorus, and Iodine. The Metals.

Experimental illustrations of the combinations of the Elements, including the phenomena of combustion, the decomposition and recomposition of water, &c.

Acids. Metallic Oxides. Earths. Alkalies. Neutralization. Salts. Solution. Crystallization. Precipitates.

LECTURE VI.
ORGANIC CHEMISTRY.

Difference between Organic and Inorganic bodies.— How to Analyse Flour, Milk, &c.—Fibres, Starches, Gums, and Sugars.—Oils and Fats.—Resins, Balsams, and Essential Oils.— Fermentation. Malt, Alcohol, Ether, and Chloroform. Breadmaking.— Vegetable Acids.— The Alkaloids.— Colouring Matters.—Constituents of Eggs, Meat, and Cheese. Gelatin.—Conclusion.

LECTURE VII.
OUTLINES OF NATURAL HISTORY.

The three Kingdoms of Nature.
Purposes of Geology and Mineralogy.

System of Classification adopted by Botanists.
Divisions of the Animal Kingdom adopted by Zoologists.
Review of Patterson's Zoological Diagrams.

LECTURE VIII.
HUMAN ANATOMY AND PHYSIOLOGY.

The bony framework of our system.
Review of the chief organs and of their functions: The Brain and the Nervous System. The Organs of Motion. The Blood; its composition, its circulation, its heat producing functions. The Lungs and Respiration.

LECTURE IX.
HUMAN PHYSIOLOGY, *continued*.

Effects of Respiration in crowded dwellings. Importance of Ventilation.

Necessity of Nutrition. The organs and functions of Nutrition. The Senses: Sight, Hearing, Smell, Taste, and Touch. The Skin; its functions, and the care required for preserving their healthy action. The Nails. The Hair.

Advantages to be derived from the study of Physiology, and manner in which it should be conducted.

Lectures on the following subjects are in preparation:—

DWELLINGS AS THEY SHOULD BE, AND THE ART OF CONSTRUCTING THEM.—BUILDING MATERIALS.—FIXTURES, FURNITURE AND HOUSEHOLD UTENSILS. — TEXTILE MATERIALS, FABRICS, DRESS.—FOOD.— WARMING, LIGHTING, AND CLEANING.—PUBLIC AND PERSONAL HYGIENE.—SAFETY FROM INJURY, AND MEANS OF RELIEF.—HOUSEHOLD MANAGEMENT, &C.

APPENDIX No. III.
(See Page 102)

NOTES CONCERNING THE FORMATION OF AN EAST LONDON MUSEUM.

N.B.—These suggestions were submitted to my friend Mr. ANTONIO BRADY, some years ago, when the scheme for which he has since succeeded in obtaining valuable Government Aid, was proposed to be realized by local subscriptions. My motive for now appending them to the present Memorandum is that they serve to elucidate my ideas as to the manner in which Science and Art Collections may be moulded into special forms for special purposes, and made up, together with appropriate accessories, into popular Institutions calculated to attract, instruct and humanize. The scale of development which I had contemplated for East London will suggest that which might suit a Maritime City, such as Liverpool, Bristol, Newcastle or Hull, or with certain alterations, an inland one such as Manchester, Leeds, or Birmingham, subject to due allowance for existing Institutions.

GENERAL REMARKS.

"The rich must not be content to do some amount of good with their money; they must do their very best, carefully weighing the several claims on their benevolence, and earnestly devoting their time, their intelligence, and their means, in those directions where they may hope to achieve the greatest amount of practical and permanent good."

"Any large sums expended in the formation of an East London Museum on any other principles than these, would give rise to legitimate complaints that they are diverted from more urgent appeals of public and private charity, in a district where much poverty and suffering exist. We must therefore consider how we can make a Museum a means of practical, incontestable, and paramount benefit to the million."

"For this purpose we must measure out sparingly and methodically our limited funds and space. We must carefully adapt our means of instruction to the capacities and requirements of the populations which we aspire to instruct. We must equally avoid admitting things which can only be appreciated by men of science or erudition, and mere curiosities without educational meaning. We must refuse articles of which the interest is great, if their bulk is still greater. We must take care not to be seduced by the liberality of our friends, to the acceptance of whole collections, of which a portion only is suited to our purpose, and we must refrain from indulging in the development of any one department beyond its due proportion to the rest."

"The course of our future operations will unavoidably be to a certain extent influenced by circumstances; but we must endeavour to keep them essentially true to their purpose, and to make them systematically tend to the accomplishment of a preconcerted and well considered programme of popular instruction. Carefully selected, well classified, and fully labelled Collections, suited for illustrating appropriate courses of Lectures, should afford to the industrious classes a ready means of acquiring that knowledge of general and technical Science, and that insight into the refinements of Art, which may best qualify them for their present or intended trades. We must seek to bring together in a popular and attractive form, all that is required for teaching the rich man how to help his less fortunate neighbour, and the poor man how to help himself and his family. We must try so to organize certain other departments, that they may present an unceasing succession of novelties, and we must in general do all that can consistently and unobjectionably be done, to render our Museum a favourite popular resort. Knowledge will not be sought unless it be made attractive, and an establishment of this kind, if not used, is not useful."

OUTLINES OF AN EAST LONDON MUSEUM.

"N.B.—The illustrations of each department, besides being as far as practicable explained by instructional labels in suitable type, will be at certain times orally explained by competent Demonstrators; moreover they will, as far as possible, be so arranged as to be easily conveyed to the Lecture Hall or Class Rooms."

A—GALLERY OF DOMESTIC AND SANITARY ECONOMY.

"On the plan of the Twickenham Economic Museum, with the omission of certain parts of Class VII (Education) which will be enlarged to form Special Galleries."

B—GALLERY OF POPULAR SCIENCE.

"Elementary Educational Illustrations of the various branches of Natural History and Philosophical Science, specially selected with a view to the Science Lectures and Classes."

C—GALLERY OF INDUSTRY.

"Technological Illustrations for the furtherance of industrial knowledge and ability, in connection with the Society of Arts' Movement for improving the technical training of Artisans."

D—GALLERY OF COMMERCE.

"Forming a convenient repertory of Commercial Samples, especially foreign and colonial, with a constant supply of appropriate information."

E—Gallery of Inventions.

"For the temporary exhibition of new and interesting inventions and improvements, which will be explained by competent Demonstrators, as at the Polytechnic Institution."

F—Gallery of Art.

"Intended to chasten and develop public taste, and to raise the standard of Art Workmanship. A School of Art to be annexed."

G—Gallery of Music.

"Consisting of a Music Hall with Class Rooms for instrumental and choral instruction and practice. Along the walls will be disposed illustrations of the History of Music, and of the several systems of Notation and Tuition used in various countries, cheap and good Instruments for the Million, &c."

H—Gallery of Literature.

"A Library of Reference for the various branches of information represented by the several Galleries. A Library of Pure Literature with arrangements for circulation. A popular Reading Room on the most approved Plan."

Miscellaneous.

"Board Room, and Office accommodation. A Lecture Hall to contain   persons, with facilities for rapid egress in an emergency. Class Rooms for popular branches of study. Accommodation for recitation and discussion meetings, &c. Facilities for every kind of unobjectionable Recreation. Facilities for indoor exercise.* A well organized Refreshment Department, in proximity to the Recreation Department and the Reading Room."

"Lavatories and other sanitary conveniences."

"Portraits of men distinguished in the various specialities to which the building is devoted, and of Working Men who by their honorable industry have raised themselves to eminence, will be distributed wherever space is appropriately available, as also other devices calculated to rouse the mind to useful activity, or to diffuse in an enlightened and conciliatory spirit, the precepts of Christian morality."

P.S.—Since the foregoing was written, I have prepared the outlines of a Collection of objects connected with EMIGRATION, designed either to form a Gallery of the East London Museum, or to be developed separately on a larger scale, so as to form a MUSEUM OF EMIGRATION. It would comprise whatever is best calculated to afford reliable guidance, and useful hints, respecting :—firstly, the choice of the future Fatherland ; secondly, the Exodus ; thirdly, the new Home ; fourthly, instruction and recreation for self and the rising family.

* Should space allow, arrangements would be made for out-door games and exercises, drilling, &c.

APPENDIX No. IV.

(See Page 102)

SUGGESTIONS FOR A DISTRICT MUSEUM OF SCIENCE, ART, AND INDUSTRY.

(*These notes were originally prepared with a view to the requirements of South London, and might meet those of a provincial Town.*)

Of paramount importance to Working Men are :—

1st. A thorough knowledge of their respective trades or occupations.

2nd. That insight into the *rationale* of "Common Things," which may enable them to turn their earnings to the best account, and to secure, as far as possible, Health and Comfort for themselves and their families.

In order to meet in some measure these requirements among the industrial population of the South of London, it is proposed to form an Institution, which, limited at first according to available means, may ultimately unite the following advantages :—

a. A large Lecture Hall, and Rooms for Class Instruction.

b. A School of Design, in connection with the South Kensington Science and Art Department.

c. A permanent Collection of Paintings, Drawings, Casts, and other objects appertaining either to high art, or to decorative art, and carefully selected with a view to raise the standard of taste, and to facilitate the progress of art workmanship.

d. Arrangements for the safe display of Works of Art, lent for limited periods.

e. A collection of apparatus, specimens, diagrams, &c., to serve for Lectures and Classes on the Elements of Physics, Chemistry, Natural History, and Physiology, as far as they are required for understanding the Laws of Health and Comfort, or as a foundation for Technical knowledge.

f. A Museum of Domestic, Sanitary and Industrial Economy, designed to unite, as far as possible, whatever may tend by visual instruction, to improve the condition of the Working Classes, in their workshops or in their homes, giving an insight into industrial materials and processes, and a thorough acquaintance with the resources of domestic life.

g. A library, to include a selection of publications relating to Elementary Science, and Domestic and Sanitary Economy, as well as the best English and foreign Handbooks of Trades.

Every exertion within the means of the Institution would be made to supplement by gratuitous or cheap instruction, whatever may be at present defective in the system of Apprenticeship, and to afford a sound scientific foundation to any who may be desirous of thoroughly studying their respective callings.

Special facilities would be provided for females wishing to perfect themselves in the various occupations suited to their sex, and especially in a knowledge of all that appertains to Domestic Economy, including the management of Children and the care of the Sick.

Music, and other eligible recreations would be duly promoted, and if possible, facilities afforded for gymnastic exercises under shelter.

APPENDIX No. V.

SUPPLEMENTARY NOTES.

The following miscellaneous Notes may serve to indicate by way of example, the kind of measures which I should wish to see undertaken in connection with the general educational movement now in progress.

Beginning with the desiderata of the higher order, I will borrow from a letter addressed to a Friend a few months since, the following suggestions :—

"PANATHENÆUM OF SCIENCE AND ART."

"The whole available portion of the space surrounding the Horticultural Grounds at South Kensington, (beyond that required for the annual International Exhibitions,) might be marked out for receiving by successive instalments, the several parts of a vast *Panathenæum* of Science and Art, fully deserving by its style of Architecture, by its various appropriations, and by its connection with the conceptions of the late Prince Consort, to be called a National Edifice. Among those appropriations I will merely mention the following:—(a.) All that is required for giving the last finish to the highest class of instruction in the *Fine Arts*, short of a journey to Rome. *Studios* to be provided on attractive terms for Artists of acknowledged eminence, who would enjoy the advantages of a refined artistic club, and exercise through their coalition a vast amount of stimulating and guiding influence in the artistic world. (b.) A complete *Conservatoire de Musique*, contiguous to the Albert Hall which would serve for occasional public performances by the élèves. (c.) A scientific University, uniting the highest aspirations with the best conveniences of the most renowned foreign scientific Establishments. The Courses given and discussion Meetings held there, would embrace the most theoretical points of scientific research, and satisfy the wishes of those who study Science for its own sake; but care should also be taken to meet every practical requirement of advanced students desiring to become first-class Engineers, Managers of Mines or Factories, &c."

The following is an enumeration of Programmes and other suggestive Papers, which I should be happy to place at the disposal of any Members of our Council who might be interested in promoting the objects to which they refer.

HYGIENE IN THE MEDICAL CURRICULUM.

a. Correspondence and Memoranda (chiefly dated 1866) respecting proposals for instituting at University College a Course of twenty-five Lectures on Public, Domestic, and Personal Hygiene. The accomplishment of this project was checked by the unfortunate illness of my talented Friend Dr. G. HARLEY.

b. Papers (dated end of 1868 and beginning of 1869) relating to a proposal for special and tentative PRIZE EXAMINATIONS IN HYGIENE at University College.* These papers include materials for a detailed Syllabus, from the introduction to which is borrowed the following classification of subjects:—

Firstly, Measures in reference to endemic and epidemic diseases; improvements in the drainage, water supply, and aeration of populous localities, and other sanitary desiderata of a Public nature, which Medical Men may have opportunities for promoting as Officers of Health, Inspectors of Sanitary Establishments, Advisers of Vestries and other Local authorities, Witnesses before Parliamentary Committees, and otherwise.

Secondly, The application of scientific principles to the design, construction and contrivance of DWELLINGS in Town and Country, and to those Departments of DOMESTIC ECONOMY which have a direct bearing on the Health and Comfort of the Inmates.

Thirdly, Matters of PERSONAL HYGIENE, in which both youth and adults require a friendly guidance, and in which Parents in particular should be taught to be the Instructors of their Children.

Fourthly, The accidents or gradual injuries to Health which attach to various Trades and Occupations.

The adoption of HYGIENE in this Country as a standard element in the official Curriculum of Medical Studies, will have a very great influence in promoting its diffusion as a part of general education throughout the Country, and this adds to the satisfaction with which I see a prospect of its recognition in an improved system of Medical Examinations, now under the serious consideration of Government.

The benefit which would result to the Public in general, and

* The examination scheme was superseded through the important determination happily arrived at by the Council of the College, to institute a regular chair of Hygiene.

to the Working Classes in particular, from an increased attention to Hygiene on the part of Medical Men, is easy to imagine. Even if sound notions as to the causes and prevention of bodily suffering should become general through the adoption of educational measures more or less similar to those proposed in this Memorandum, yet still whenever the collective efforts of the Volunteers of Health are required for overcoming a common enemy, they must be led to victory by MEDICAL OFFICERS.

c. Outlines of Classification for a comprehensive MANUAL OF HYGIENE.—I was led to prepare the paper here referred to, through the wants and difficulties which arose in attempting to make provision of suitable Class Books, for the earnest and methodical study of Hygiene mentioned above. There is no absolute lack of valuable information, but it is scattered in different Books and Periodicals, not all in English, and not always free from the admixture of uncertain theories, unsifted data, and interested or prejudiced appreciations. My wish is to see brought out under the auspices of a Body affording the most reliable guarantees, and in a style of completeness which I should be happy to promote by pecuniary advances, a Standard Work jointly elaborated by first-rate Men, and which in matters of Public, Domestic, Personal, and Technical Hygiene, may not only be a Cyclopædia of reference for Officers of Health, and other advanced Hygienists, and a normal Text Book for Medical Students, but may also serve as an acknowledged source whence the instruction required by the various classes of the community may be made to flow to them, at various levels, through appropriate channels. At the same time this digest of the Laws of Health should recommend itself to the general reader, by a clear and easy style, free from all unnecessary technicalities; but he would be expected to have a good previous acquaintance with Natural Science.

Hygiene when taken in the above comprehensive sense, extends over by far the greater portion of the ground which I have described in this Memorandum as occupied by the Science of Common Life; but in the ordinary acceptance of the word Hygiene, certain considerations have too often been omitted, without which Hygienic knowledge cannot satisfactorily be applied to the benefit of the Million. Matters savouring of Domestic Economy have been looked down upon by those who could have brought most Science to bear on them, and neither the means nor the inclinations of the class of people to be benefited, have always been sufficiently attended to. Yet it is of little avail to recommend an article of Food on account of its nutritiveness, unless it be palatable, or a mattress, be it ever so sanitary, unless it be comfortable to lie on ; and if they are so far right, the further question has still to be considered,—are they cheap enough for general use ? Again it is in most cases necessary to look into the technology of the things used in Daily Life, in order to understand whether and why that use is to be recommended, and also whether and how adulteration and fraud are apt to creep in to the detriment of health.

It is through considerations like these, that I have been induced to adopt the term PRACTICAL BIONOMY, as coupling with the ordinary range of hygienic knowledge, all that might most effectively tend to make it the purveyor of COMFORT as well as of HEALTH.

HEALTH AND COMFORT FOR SEAMEN.

Such is the title of two Lectures in an advanced state, besides which I have in hand the Programme of a more extended Course for Commanders and Officers of Vessels not having a Surgeon.

Though I have endeavoured to make the scope of my Popular Course embrace as large a proportion as possible of the Working Class, yet there are certain industrial denominations which must necessarily be taken by themselves in any attempt at improvement. Such is particularly the case with SAILORS. The praiseworthy endeavours of the Rev. DAN GREATOREX at that invaluable Institution, the Sailors' Home in Well Street, and among other things the trial there of one of the Lectures of my Course, have established the fact that seamen can under favourable circumstances, be induced to listen attentively to a Lecture or two even of a scientific character. But they cannot be expected to attend a Course, and I am consequently endeavouring to condense into two Lectures the most essential points of what may be called NAUTICAL HYGIENE. As regards the intended Course for Commanders and Officers in the Merchant Service, I need scarcely say how very important it is that they should know how Health and Comfort may best be promoted by them under the peculiar circumstances of nautical life, and in particular how vitally urgent it is that those who in the absence of a Surgeon are entrusted with the regulation Medicine Chest, should have some kind of regular training to a judicious use of it. Nothing seems more natural than that Candidates for Captains' and First Mates' Certificates, should be required to give proof of such training in their Examination, and it is my object, *firstly*, to show that the amount of knowledge required, together with the scientific foundation on which it must indispensably rest, may by a careful selection and adaptation, be brought quite within a manageable compass; and *secondly,* to endeavour to raise in the proper quarter, on the one hand a determination to require strictly the knowledge in question, and on the other hand a willingness to afford on the most inviting terms, convenient opportunities for acquiring it.

I should be equally happy to place in the hands of any friends interested in these matters, a detailed Syllabus of two Lectures on FRESH AIR and PURE WATER, specially designed for the poorer class of Irish Labourers in London; and a sketch of two Lectures on "Cholera and its Prevention;" also a paper on Theatrical Representations as a means for raising instead of lowering the tone of moral sentiment among the People; and other manuscripts on miscellaneous devices for improving the condition of the Working Classes.

I must apologize for having mixed in this Memorandum, subjects so divergent in character, that their juxtaposition may at first sight seem incongruous; but I trust they will all be found to be connected with each other, and with the movement for the promotion of Technical Instruction, in which our Society is engaged, by a fundamental principle; that of recognizing Science as the best ally to Christian Civilization; as an agency designed by Providence to foster the spirit of industry and orderly progress, to render men's ways prosperous, and to make their homes the abodes of peace, well-being and thankfulness.

www.ingramcontent.com/pod-product-compliance
Lightning Source LLC
Chambersburg PA
CBHW020101170426
43199CB00009B/361